Redefining Land Management in North American Literature and Culture

Exploring the ongoing histories of human-centered ecosystem management in the lands and waters that comprise what is now known as North America, this book tracks the diverse ways in which human-environmental relations have been presented across different forms of media.

Including literature, film, visual arts, performing arts, park interpretive materials, and more, the book analyzes the many ways in which human-environmental relations have been articulated, experienced, understood, represented, and regulated. The collection encourages a reimagining of what it means to manage environmental elements beyond the limits of utility, commodification, and control. To that end, the contributions interrogate the concept of "management" itself, arguing instead for a more expansive conception of reciprocal relations among humans and the more-than-human world. These relations encompass many kinds of human-environmental engagements, ranging from constructing dams and using prairie soils for paint pigments to gathering pollen and cultivating grass for memorial sites. In particular, the book highlights the potential in Indigenous ecological approaches to living with lands and waters, underscoring the need to lovingly engage with the world that sustains us. In the spirit of inquiry and reorientation, this interdisciplinary book begins with a "Keywords" section. Entries for key terms, such as wilderness, allotment, reclamation, fugitive sand, and reciprocity, detail the word's history, outline its role in various discourses, and/or suggest possibilities for future research.

This book will be of value to a wide range of readers, including researchers and students in environmental humanities, literature, ecocritical theory, history, Indigenous studies, legal studies, natural resources, sociology, environmental studies, and anthropology.

Jada Ach is Associate Teaching Professor of Applied Humanities and Interdisciplinary Studies at Arizona State University where she teaches courses in literature, film, and environmental humanities. Her research combines insights from literary studies, the desert humanities, and land management history. She is the author of *Sand, Water, Salt: Managing the Elements in Literature of the American West, 1880–1925* (Texas Tech University Press, 2021) and co-editor of *Reading Aridity in Western American Literature* (Lexington Books, 2020).

Kristen Brown is Assistant Professor of English at Northern State University in South Dakota, where she teaches environmentally themed composition and literature courses. Her teaching and research center Indigenous authors and perspectives with attention to the sensory—especially sonic—contours of being and belonging. In addition to her chapter on pedagogical approaches to the works of Charles Eastman/Ohiyesa (Santee Dakota) in the award-winning collection *Race in the Multiethnic Literature Classroom* (University of Illinois, 2024), her articles appear in *Western American Literature* and *Resonance: The Journal of Sound and Culture*.

Routledge Environmental Humanities

Series editors: Scott Slovic *(University of Idaho, USA)*
Joni Adamson *(Arizona State University, USA)*
Yuki Masami *(Aoyama Gakuin University, Japan)*

The *Routledge Environmental Humanities* series is an original and inspiring venture recognising that today's world agricultural and water crises, ocean pollution and resource depletion, global warming from greenhouse gases, urban sprawl, overpopulation, food insecurity and environmental justice are all *crises of culture.*

The reality of understanding and finding adaptive solutions to our present and future environmental challenges has shifted the epicenter of environmental studies away from an exclusively scientific and technological framework to one that depends on the human-focused disciplines and ideas of the humanities and allied social sciences.

We thus welcome book proposals from all humanities and social sciences disciplines for an inclusive and interdisciplinary series. We favour manuscripts aimed at an international readership and written in a lively and accessible style. The readership comprises scholars and students from the humanities and social sciences and thoughtful readers concerned about the human dimensions of environmental change.

For more information about this series, please visit: www.routledge.com/Routledge-Environmental-Humanities/book-series/REH

Redefining Land Management in North American Literature and Culture

From Resource to Reciprocity

Edited by Jada Ach and Kristen Brown

Routledge
Taylor & Francis Group
LONDON AND NEW YORK

earthscan
from Routledge

First published 2026
by Routledge
4 Park Square, Milton Park, Abingdon, Oxon OX14 4RN

and by Routledge
605 Third Avenue, New York, NY 10158

Routledge is an imprint of the Taylor & Francis Group, an informa business

British Library Cataloguing-in-Publication Data
A catalogue record for this book is available from the British Library

Library of Congress Cataloging-in-Publication Data
Names: Ach, Jada, 1980– editor | Brown, Kristen, R. editor
Title: Redefining land management in North American literature and culture : from resource to reciprocity / edited by Jada Ach and Kristen Brown.
Description: London ; New York : Routledge, 2026. | Series: Routledge environmental humanities, 2691-9591 | Includes bibliographical references and index.
Identifiers: LCCN 2026008468 (print) | LCCN 2026008469 (ebook) | ISBN 9781032946757 hardback | ISBN 9781032946764 paperback | ISBN 9781003581185 ebook
Subjects: LCSH: Nature in literature | Land use in literature | American literature—History and criticism | Canadian literature—History and criticism | Human ecology in literature | Ecocriticism
Classification: LCC PS163 .R45 2026 (print) | LCC PS163 (ebook)
LC record available at https://lccn.loc.gov/2026008468
LC ebook record available at https://lccn.loc.gov/2026008469

ISBN: 978-1-032-94675-7 (hbk)
ISBN: 978-1-041-33287-9 (pbk)
ISBN: 978-1-003-58118-5 (ebk)

DOI: 10.4324/9781003581185

Typeset in Sabon
by Apex CoVantage, LLC

Dedicated to Juniper and Winona, with love.

Contents

Contributors

Jada Ach
Arizona State University
US

Lydia Cagluaq Agnus
National Park Service
US

Joshua T. Anderson
Flyway Foundation and *Common Ground: A Prairie Podcast*
US

Surabhi Balachander
Oregon State University
US

Michael R. Barnes
University of Minnesota
US

Kristen Brown
Northern State University
US

Cari M. Carpenter
West Virginia University
US

Jessica Cory
Appalachian State University
US

Andrea Everett
MatriARC PROJECTion, LLC
US

Lisa Fink
Michigan State University
US

Quinn Grover
Brigham Young University—Idaho
US

Robert Gunn
University of Texas—El Paso
US

Amy T. Hamilton
Northern Michigan University
US

Mika Kennedy
Ithaca College
US

Sarah Jane Kerwin
Oberlin College
US

Alex Mares
Hueco Tanks State Park and Historic Site
US

Rafael A. Martínez
Arizona State University
US

Hannah Muhlfelder
Boston University
US

Sreya Ann Oommen
Arizona State University
US

Eytan Pol
Independent Researcher
US

Jonathan Radocay
University of Washington
US

Gary Reger
Trinity College
US

Dominique Aurilla Vargas
University of Texas at San Antonio
US

Joseph Gazing Wolf
Arizona State University and Heritage Lands Collective
US

Elpida Ziavra
National and Kapodistrian University of Athens
Greece

Foreword

I lost my Grandfathers, Fawzy and Victor, way too soon, a price one pays for being born poor. They were both deeply loving men. Their spirit, their breath, was forged in the fire of Indigenous, land-birthed communities from opposite sides of Mother Earth.

I remember Gedo (Grandfather) Fawzy's loving embraces and sweet kindnesses. He would put me on his work-worn and tired shoulders—emaciated by a lack of adequate nutrition—and take me to his machine shop. His ancestors had been among the many Nubian-Egyptians who were forcibly removed from their homelands with the construction of the Aswan Dam. The Dam allowed the British Empire to control the life-giving waters of the Nile to feed their appetite for cotton, while destroying the soils, plants, wildlife, and human communities that relied on the Nile's natural flow. The people, being adaptable—like the very waters that colonizers were attempting to control—learned to survive through their ancient, skilled means. Gedo Fawzy, having no formal schooling, was producing aircraft parts for well-known international companies in his meager machine shop. Like Teta (Grandmother) Wadi'a, who made miracles in her nonexistent mud-floor kitchen, Gedo Fawzy had a life-giving touch—what Daniel Wildcat, Yuchi, and Muscogee Elder would call life-enhancement.

I cannot imagine the daily stress Gedo carried in his heart, providing for a large family in a state of extreme poverty, landlessness, and colonial humiliation. Despite this, his heart was full enough to turn to me whenever I was feeling anxious about the difficult life I was living. He would wrap one of his arms around me and hold me close to him, tease me or kiss me, and whisper something funny in my ear. His calming whispers mocked the absurdity of vapid humans and their institutions, which were the primary source of my tribulations. This included the insane behavior of "leaders" in our communities, the Euro-puppet Egyptian government, and the ever-present European exploiters of "natural resources." When we were alone, he would tell me incredible stories about our ancient ones and their sacral intersecting relationships with plant and animal kin, water and wind, and language and wisdom. Though he descended from the people who created writing, the paper, the library, the research university, and the scientific method, Gedo was a quiet man, living in a mindful, steady spirit. He was like water, always moving toward a state of rest.

Lalá (Grandfather) Victor was the Húŋkpapȟa Lakȟóta version of Gedo Fawzy. As a child, he was taken from his tiošpaye (extended family) to a boarding school. There, an attempted murder of any element of his Lakȟóta culture ensued. Like a dam, genocidal colonial culture obstructed the flow of his life into ever-deepening ties to his sacred kin (Mitákuye Oyás'iŋ—"All my relations"—a reference to the interconnectedness of all life forces). Despite the ephemeral capitalist darkness that threw his once fiercely independent people into dependency and subjugation, Lalá Victor illegally kept the Sundance, maintained his language, practiced his prayers, and was nearly killed for it. Though he had never seen Tȟatȟáŋka Oyáte (the Buffalo Nation) until he was an older adult, because they were nearly massacred into extinction, it was their steady presence that made him endure. He remembered the songs of Tȟatȟáŋka, our shared origin story, their guidance for our people, and their centrality to our sacred relationship with the Earth. Most of all, he kept memory of their breath.

Niyá, the breath or spirit that animates all beings, connects our life in a very direct way to every being on Earth and beyond. The Niyá shared between Tȟatȟáŋka and Lakȟóta is so direct that Lalá said if the buffalo are not doing well, we are not doing well. We are one, the same people and breath, the Pte Oyáte, expressed in differing forms, some as human and some as buffalo. Lalá Victor taught me all about the buffalo spirit: their movement that is deeply in sync with the rhythms of the land, their matriarchal social organization, their quiet and mindful nature, and their free and indefatigable spirit. Their eyes pierce the soul, and their breath fills it with life. Like Tȟatȟáŋka, Lalá was steady, quiet, mindful, and powerful.

"We are the forces of nature defending herself," the Earth cries out. Like the life-giving waters of the Nile, Mni Wičoni ("water is life"), Gedo Fawzy embodied a steady movement, sometimes in upheaval against oppression and collective stupidity, more often in soft movement around obstacles, always toward restful depth. The life-giving breath of Tȟatȟáŋka animated Lalá Victor, a spirit that made him an outlaw in the presence of colonial mindlessness. He breathed their gentle compassion, their fierce defense of fellow buffalo, their silliness, their life-enhancing movement, their honoring of matriarchs, and their unrestrained spirit.

Love does not triumph nor prevail! The spirit that animates us all teaches us that love is a small, ironic point of light in an otherwise dark world (from W.H. Auden's poem "*September 1, 1939*"). It never flinches, never stops moving, never ceases giving its light, and never succumbs to Thanatos (Greek for "death"). The seemingly monstrous yet ephemeral agents of Thanatos attack it with tanks, drones, guns, and bombs. They starve it and genocide it. They build dams to control it, fences and cages to restrain it, and "sustainable development" to replace it. They smear it with vicious lies, alternative facts, and narcissistic droning. Yet, the absurd worshipers of death always stand impotent in their struggle to extinguish love, a chasing after the wind ("Hevel hevelim" from Qohelet—Hebrew for "Absurdity of absurdities").

Love endures! Love is the gentle spirit that animates life and gives it meaning. Its powerlessness is its immortality. Love is the resurrection!

> Human history is not the battle of good struggling to overcome evil. It is a battle fought by a great evil struggling to crush a small kernel of human kindness. But if what is human in human beings has not been destroyed even now, then evil will never conquer.
>
> — Vasily Grossman, *Life and Fate*

Joseph Gazing Wolf

Acknowledgments

A book is always a collaborative effort, and this one is no exception. We first want to thank our contributors, without whom this collection would not have been possible. We learned so much from each of them, and we're eager to see what kinds of meaningful impacts their work will have on the communities that extend beyond publication.

We also want to express our gratitude to the Association for the Study of Literature and Environment, which allowed Jada to organize a panel focused on land management in North American Literature at the 2023 conference in Portland, Oregon. The papers shared on that Western-Literature-Association-sponsored panel seeded the idea for this book.

Both of us feel deep gratitude for the Western Literature Association. Their conferences have consistently nurtured spaces for sharing ideas and supporting graduate work, which is how we first engaged with the organization and found the confidence to flex our voices. It is deeply satisfying to acknowledge them here, in the early stages of our career, and we look forward to supporting graduate students in turn. We are excited to see what the next fifty years of the organization will bring!

Our institutions—Arizona State University and Northern State University—deserve many thanks as well. In addition to their support in attending conferences, our colleagues have participated in writing workshops and offered encouragement along the way.

We appreciate the editors and peer reviewers for Routledge's *Environmental Humanities Series*, who sensed the value in this collection and provided invaluable feedback. They too saw the possibilities in this project reaching beyond the boundaries of academia.

Many deep and heartfelt thanks are due to our families for their love and patience. Our love for our daughters, Winona (Jada) and Juniper (Kristen), is the primary motivator in our celebrations of life and struggles for justice. The future belongs to them.

Introduction

Resourcing Love in Land Management

Jada Ach and Kristen Brown

Crisis and Collaboration: An Invitation

This collection emerges from experiences and observations that created guiding memories in our lives, the earliest occurring long before either of us decided to enter academia. Many of us can recall a moment in childhood that revealed the extent to which humans have devastatingly impacted the planet's ecosystems. One of those memories, for me (Kristen), was the frequent closing of beaches along the shores of Lake Erie in the late 1980s due to wastewater pollution. When I learned that the nearby Cuyahoga River had caught fire, multiple times, I felt fear, anger, and distrust. For me (Jada), that realization arrived when a local park in my hometown put up signs warning individuals to avoid fishing or swimming in the small body of water once named Conservation Lake. The rumor was that the waters had become polluted from old car batteries, but in reality, years of industrial runoff had heavily contaminated the spring-fed swimming hole, making it unsafe. An event that we shared, however, was the "Thousand-Year Flood" in early October 2015, while we were both starting our third year as PhD students at the University of South Carolina. As Hurricane Joaquin lingered off the southeastern coast for four days, the constant rain was too much for the city's infrastructure to bear. Bridges burst, dams failed, interstates flooded, and coffins were unearthed. Each time our cellphones buzzed with a new emergency alert, we braced for the worst.

We exchanged nervous texts during the flood and its aftermath. Our text messages provide not only a kind of story in themselves, a story of our friendship, of jokes, of joy, of heartbreak—but also a story of how crisis connects us, how we can love fiercely in our work toward repair and renewal. Since that flood, each of us has brought a child into this world, even as we struggle to manage the fear of proliferating environmental crises. The word "manage," with its modern associations of hierarchical, top-down control, suggests that we can contain what often cannot be contained. It often forecloses the opportunity to intimately witness, to live with and through. It stifles our ability to accept and embrace that which is unknown—or perhaps even unknowable, unmanageable—to us. And in a settler-colonial context, when

DOI: 10.4324/9781003581185-1

those in power fear that their management efforts might fail, they have often chosen coercion over dialogue, violence over collaboration.

In this collection, we've chosen dialogue and collaboration as our guideposts. As land management is, undeniably, a collaborative act that occurs between diverse human and nonhuman stakeholders, we decided early in the process that collaboration should also be central to the collection's themes and structure. After all, if we're calling for a vision of land management that is transdisciplinary, community-centered, and dedicated to listening and witnessing, then we wanted those values to be reflected in the very bones of the collection. To that end, this project includes coauthored keyword entries and chapters, as well as interdisciplinary and transdisciplinary collaborations. We also asked all of our contributors to engage with and reference each other's work throughout the process, and you'll see those citational collaborations happening across nearly all of the contributions. We look forward to keeping these conversations going in future projects, panels, and gatherings.

We invite you, the reader, into this collaboration and hope that the offerings here encourage you to reflect on the memories that shape how you attend and attune to a more-than-human world. We also hope you might consider how reconfigurations of land management—a term we intentionally forward as worthy of interrogation, as something in need of reimagining—might guide your engagements with the people and places that collectively create the way we live, work, play, and love.

From Indigenous Stewardship to Colonial Control

Well before colonial contact in what is now known as the Americas, sophisticated human systems developed over thousands of years of place-based trial and error aligned with diverse ecological communities—from farming in Mesoamerica to fish preservation in the Pacific Northwest to marine mammal management in the Arctic. These names—Mesoamerica, Pacific Northwest, Arctic—are themselves relatively recent when considered with these deeper histories. We might trace a trajectory of Euro-Western power and control from the imperial logics of the 1493 Doctrine of Discovery to their codification in *Johnson v. M'Intosh* (1823), a ruling that denied Indigenousproperty rights by establishing that the original inhabitants held only a "right of occupancy" while the "discovering" European nation held ultimate title—a legal framework that seemingly justified the dispossession and erasure of Indigenous Peoples. This case, along with the two cases that followed—*Cherokee Nation v. Georgia* (1831), which redefined sovereign Indigenous Nations as "domestic dependent nations" in a ward-guardian paradigm, and *Worcester v. Georgia* (1832), which affirmed tribal sovereignty while simultaneously consolidating exclusive federal authority over Indian affairs—formed what became known as the Marshall Trilogy. Together, these rulings subjected Indigenous Nations to the plenary power of an eager-to-expand settler-colonial nation-state. Ironically couched in Enlightenment-era humanist ideals and driven by a civilization discourse buttressed by the power of church and state, the US legal system imposed a paternalistic framework that systematically undermined multigenerational Indigenous relationships with homelands.

The violent removals and enforcement of a reservation system in the decades that followed further disrupted traditional ecological relations: deforestation and soil-stripping monocultures were soon followed by barbed-wire fences and mining. To expedite this new approach to land relations, the Allotment Act in 1887—legislation that rendered the United States a "perverted checkerboard," to use an apt figure from Standing Rock author-activist Vine Deloria, Jr.—was paired with the assimilationist engine of the Indian Boarding Schools which sought to erase and replace.[1] In his recent report on the Indian Boarding School System, Assistant Secretary of the Interior Bryan Newland (Obijwe) underscored "the twin policy of Indian territorial dispossession and Indian assimilation," an important reminder that cultural suppression and environmental exploitation were, and are, mutually reinforcing.[2] The violent displacement and cultural disruption that followed these legal foundations threatened not only human communities but also the lands and waters themselves, the more-than-human ecological communities they sustained through diverse place-based knowledges woven into dynamic sophisticated systems.

The United States and Canada shared playbooks for the accumulation of lands and eradication of Indigenous lifeways. The legacies of these attempted erasures of Traditional Knowledge Systems—and the imposition of top-down supremacist regimes of containment and control—continue to reverberate. The criminalization of Indigenous cultural burns, and the catastrophic wildfires that have followed, provides but one example. Our contributors consider several other examples, tracing how the dominant discourses of management—allotment, assimilation, borders, reclamation, wilderness are registered, critiqued, and challenged in cultural and literary expressions. Collectively, the offerings here unsettle the relatively recent assumption that ecological communities are merely resources to be managed. Instead, they resituate land management within a more expansive understanding that invites us to defamiliarize settler-colonial, mechanistic perspectives, and practices, to consider the need for care and repair in tending the lands and waters that sustain us. In the chapter that concludes this collection, former lead ranger at Hueco Tanks Alex Mares notes, "we don't manage a damn thing." He insists, instead, that "we need to manage ourselves."

Redefining Land Management: From Resource to Reciprocity

"Land management" is a vexed term often aligned with masculinist, settler-colonial priorities related to accumulation, classification, and control. In the United States, scientific land management, as we know it today, emerged in the late nineteenth century and often privileged Western, economic, and imperialistic ways of knowing the land. From this view, the environment exists *out there*, beyond the (white-male human) body, with untapped resources-in-waiting. Resources are central to the settler-colonial land management project; in many ways, the project exists to protect, use, or profit from the diverse natural resources that come to be valued in the global economy.

As a term that many of the chapters in this collection challenge or rewrite, "resource" has the effect of rhetorically deadening and abstracting environmental matter, robbing it of its life/liveliness and inhibiting it from connecting generatively with other biological, cultural, and spiritual systems. Furthermore, "resource" forces us to imagine the environment as a collection of severed pieces, some of which are more valuable—and, hence, more extractable—than others. Finally, the centrality of "resource" in the land management project prevents us from meaningfully connecting with the present moment, the place where intimate engagements, listening, and community-building might occur. Instead, the concept demands that we dwell on future scenarios of scarcity and wealth accumulation instead of being accountable for and loving what is already here, in the now. *Will I have enough?* is the anxious question "resource" asks us, and not *How can I be in better relations?*. Therefore, in this collection, we choose to *resource love*, to refuse the colonial logic coded in "resource" and transform it into a verb that reminds us of the return and repetition in our own sourcing, our own ability and boundless capacity to love, and "to rise" (Latin *surgere* for source) to the moment that we find ourselves in. This semantic adaptation of resource restores its connection to resurgence, a term Anishinaabe author-activist Leanne Betasomasake Simpson describes as "fundamentally non-hierarchal, nonexploitative, nonextractivist, and nonauthoritarian"[3]

In order to resource love and model accountability in land management, we must acknowledge the radical ways that stories invite us into relations with the land. Land management, as we aim to redefine it in this humanities-based collection, is never just about overseeing the environment. It is also, and *always,* an act of storying place—of putting into practice the diverse ways we relate to, desire, imagine, and remember our ties to the land. For anytime we engage with the land—through work, research, resting, recreating, teaching, ceremony, fishing, farming, etc.—we are both storying the land and engaging with lands already storied. And through these very acts, the process of re-storying land, or of reactivating the present by bringing remembered relations to the surface, is also possible. According to desert humanists Celina Osuna and Aidan Tynan in their collection *Storied Deserts*, acknowledging land as storied allows us to "move traditions, histories, and cultures that have been systematically subjugated or erased to the forefront of our environmental consciousness."[4] By choosing reciprocal, listening-based, community-oriented, and love-informed approaches to land management—approaches that, to borrow the words of Potawatomi botanist Robin Wall Kimmerer, "name the world as a gift" so as to feel our "membership in the web of reciprocity"—we may have the ability to re-story human-land relations that bend in the direction of justice.[5]

Love as Environmental Methodology

As many of the chapters in this collection will attest, land management is also imbricated with strong emotions like love, pride, and even fear. According to self-described "queer Black feminist love evangelist," Alexis Pauline Gumbs, "Nothing is objective. . . .[Scientists] are clearly obsessed, and most likely, like me, in love. Whether they can admit it in their publications or not."[6] For even though "public lands management frames emotion and science as being at odds with each other," as human-animal studies scholars Jennifer Britton et al. attest, one's feelings, values, and attachments always (and powerfully) inform how they choose to manage their environments.[7] In their analysis of how affect, gender, and politics inform public debates around wild horse roundups and removals in the US West, for example, Britton et al. note that purely objective, emotionless stances in these debates do not exist—even if that's how rangeland ecologists and the Bureau of Land Management (BLM) perceive of their positions. The authors remind us that "rationality and objectivity are themselves feelings," as is the love and empathy that motivate many wild horse advocates who oppose the roundups.[8] And as the horses, themselves, "experience rich inner lives and social relations rich in emotions," we begin to see that feelings have *everything* to do with land management.[9] Especially love.

For many land managers, love motivates their work—and they are proud to declare it. For example, when Cam Juárez, the Community Engagement Coordinator for Saguaro National Park, visited one of my (Jada) classes, he asked my students to raise their hands if they have ever been in love. He then said, "When you visit these special places, it's like falling in love. Whether it's a romantic partner, a newborn child, or a puppy, you want to be around and protect what you love, right?" His job as a ranger, he then explained, is to help visitors find pathways for falling in love with the park. Likewise, humanists and artists are good at bringing love to the table. Skilled at listening to the subtleties of a text, building new archives, and slowing down, humanities scholars, we argue, are necessary collaborators in imagining new land management futures. Across this collection, our contributors try to demonstrate what humanities-based participation in North American land management debates has the potential to do.

In their collection *Weaving Otherwise*, Amanda R. Tachine and Z Nicolazzo underscore the applied, relational, and transformative power of humanities methodologies: "Qualitative research methods are also action oriented (a verb), creating threads where we recognize and feel more deeply that we are in relations with life and the world around us. Nothing is solitary, and no one is singular."[10] Furthermore, Lisa Fink, environmental humanist and former postdoctoral scholar for the National Park Service's (NPS's) Alaska Regional Interpretation Team, has compellingly argued that humanities-based research in national parks can help prevent overburdening Indigenous communities with cultural interpretation projects.[11] Reiterating

Fink's disciplinary call, historian of science, Libby Robin, has argued that humanists are necessary collaborators in land management projects as "the complexity of managing places" calls for a diversity of opinions, insights, and approaches.[12] The humanities, it seems, has much to offer environmental management research, including new place-based methods, story-rich archives, and affects. The chapters in this collection build from these ongoing conversations while pointing out new collaborative pathways. In other words, they "align research with our heart, values, visions, and hopes," to use the words of Tachine and Nicolazzo, in their meditations on management.[13]

According to bell hooks in *All about Love: New Visions*, "love and abuse cannot coexist."[14] As editors, we have been inspired by the environmental possibilities contained in this seemingly simple phrase and believe that if we wish to manage better, we need to learn to love better. As many of our contributors argue, public lands can play a critical role in helping to foster the kind of love that rejects ecological abuse, neglect, abstraction, and severance. Writer and activist adrienne marie brown views love as a powerful strategy for activating social change. "Love leads us to observe in a much deeper way than any other emotion," she says—and with observation comes understanding and affiliation, a desire to remain close to the rock/river/pigeon/desert/crabapple you are observing.[15] Artists and humanists are experts at developing strategies for connecting people to place, of pursuing love as a way of knowing. One of many examples is 24th Poet Laureate Ada Limón's *You Are Here: Poetry in the Parks* project, a collaboration between Limón, the Poetry Society of America, the Library of Congress, and the NPS. The project celebrates the innumerable ways that poetry connects individuals to the natural world and has resulted in readings, site-specific poetry installations, and a collection of nature poems. "Never has it been more urgent to feel a sense of reciprocity with our environment," says Limón, "and poetry's alchemical mix of attention, silence, and rhythm gives us a reciprocal way of experiencing nature—of communing with the natural world through breath and presence."[16] In this view, poetry *is* management—it channels our breath, our attention, and our emotions into more meaningful relations with the land.

Design and Structure

This collection is structured around four parts, each of which calls on readers to reconsider the environmental possibilities of integrating humanities insights with land management policies and practices in various temporal, cultural, and geographical contexts. The continental scope of this project is intended to disrupt conceptions of "the nation" as a fixed, impermeable, and uncontested category and to, instead, acknowledge the unbound realities of dynamic and rapidly changing ecological communities. Recognizing how treaties injected imperial terminology like "nationhood" into Indigenous discourse, Scott Richard Lyons (Ojibwe/Dakota) notes that the term provides "a way to gather and organize resources that can help improve people's lives, so I say keep it around while de-essentializing it."[17] Still, we do not wish to ignore the administrative power that "the nation" has continued to wield in its various efforts to manage its ethnic, racial, and cultural identity, a power that frequently manifests as rhetorical or physical violence against those who threaten the state's conception of itself. Attending to the fraught language that both enables and constrains our shared world, we see definitional work as a fruitful entry point for unsettling some of the categories and classifications of land management.

To nourish this semantic sensitivity, Part 1, titled "The Language of Management," considers how words shape environmental thought and practice. The scope and spirit of these terms range widely. For example, Jonathan Radocay's treatment of *allotment* interrogates the devastating afterlives of federal legal language while Lisa Fink and Lydia Cagluaq Agnus provide intimate stories that illuminate the lived reality of, and animating possibilities in, *relationality*. Together, all nine keyword entries explore the cultural and legal frameworks that create and perpetuate the discourse of land management while opening a space for more inclusive perspectives and participatory language. Though certainly not a definitive lexicon, the terms included in Part 1—*badlands management* (Joshua T. Anderson), *relationality* (Lydia Cagluaq Agnus and Lisa Fink), *reclamation* (Quinn Grover), *allotment* (Jonathan Radocay), *islandness* (Sreya Ann Oommen), *aridity* (Gary Reger), *(un)documented ecologies* (Rafael A. Martínez), *fugitive sand* (Jada Ach), and *rural/wilderness* (Surabhi Balachander)—give humanists and land managers an opportunity to see how their interests and goals overlap, challenge one another, and open space for meaningful collaboration.

After considering the power of language in articulating, resisting, and reimagining land management, Part 2 interrogates the histories of claim and control in North American environmental management. "Representations of Management: Histories of Claim and Control" includes explorations of race, ethnicity, gender, and sexuality in uneven power structures that impose borders and claim ownership through state-sanctioned control. The authors in Part 2 make a case that the settler-colonial project of managing land has also always been a project of managing people. Authors Jessica Cory, Mika Kennedy, and Eytan Pol examine diverse histories and technologies—hydroelectric dam building, WWII Japanese incarceration camps, and trail rating apps—to reveal how land management practices impact human lives and the ecological systems in which they are entangled. As diverse responses to these impacts, the chapters in Part 2 offer narrative alternatives that attempt to resist land management's infrastructural, carceral, and/or technological regimes.

Part 3, "Webs of Caring Relations: Cultivating Ecorelational Literacies in the Environmental Humanities," speaks back to the top-down administrative logics of the managerial practices discussed in Part 2 by offering lateral, collaborative, and interconnected visions of land management that unsettle hierarchies of settler-colonial domination. The chapters in this section—authored by Elpida Ziavra, Dominique Aurilla Vargas, Hannah Muhlfelder, and Kristen Brown—provide ways of understanding care-centered relations with land in North American literature and culture while highlighting the potential of collaborative stewardship. To counter unilateral extractive frameworks, we propose here the concept of "ecorelational literacies"—an expanded understanding of literacy that moves beyond the alphabetic to include ecological communities themselves as texts that are not only legible but also sentient and sensible. Where settler-colonial land management has often relied on mechanistic literacies that reduce complex ecosystems to quantifiable resources, maps, and boundaries, ecorelational literacies recognize the land as a dynamic, unbounded, communicative system requiring attentive listening and reciprocal engagement.

Unlike Western literacy's exclusionary histories and modern associations with rigid educational structures, ecorelational literacy works to unmake that epistemic violence and re-story the land. Remembering her grandparents, Blackfoot scholar Rosalyn LaPier recently wrote: "Like a favorite and familiar book, the land was a text that they read to entertain and enchant their grandchildren and to share our history."[18] Well aligned with this conception of land as text, Kim et al. define ecoliteracy as the ability "to identify names, uses, and related stories" based on Traditional Ecological Knowledge (TEK).[19] The addition of *relational* reimagines ecoliteracy as not only textual, epistemological, and ideological but also sensorial, experiential, and reciprocal. This literacy is plural, collaborative, and always in-process. The practice of ecorelational literacy recognizes that we, too, are sensed and appreciated by the ecological communities to which we belong. Fire-dependent ecosystems hunger for the fire that lightning doesn't always bring. Certain plant species—like manoomin (wild rice)—thrive through Indigenous harvesting practices that have shaped them for tens of thousands of years. Pacific salmon runs are returning with the guidance of Indigenous-led management developed over many lifetimes and designed to maintain healthy populations. To practice ecorelational literacy is to understand that ecosystems can thrive through particular forms of human participation—not domination or extraction, but reciprocal care that honors both ecological agency and human responsibility within interconnected, interdependent webs of relations.

While Parts 2 and 3 evoke conceptual challenges to a normalized extraction-based environmental ethos while envisioning a more inclusive web of multispecies relations, Part 4, "Resourcing Love to Actualize Otherwise Worlds," explores applied and public practices of "resourcing love" by asking where and how we might realize more generative and loving ecological values. Contributors develop a variety of critical tools, pedagogical strategies, experiential methods, and discursive techniques in order to open spaces for communities whose ethics and policies center equitable justice, environmental attunement, good relations, and potentially unconditional love. The sites they investigate and archives they build are diverse, including a national military park, climbing memoirs, NPS management documents, a state park and historic site, cookbooks, and campus gardens. Authors include Michael R. Barnes, Amy T. Hamilton, Cari M. Carpenter, Sarah Jane Kerwin, Robert Gunn, Andrea Everett, and Alex Mares.

***

At the time of this writing, public lands in the United States face unprecedented environmental challenges under the second Trump Administration. The threats can sometimes feel difficult to absorb: renewed calls to "drill, baby, drill"; budget cuts and layoffs at national parks and forests; new proposals to lease or sell public lands; and plans to mine at sacred sites like Oak Flat and Thacker Pass. At the same time, storytelling in national parks is also under threat. Signage has begun to appear in parks urging visitors to report "any signs or other information that are negative about either past or living Americans or that fail to emphasize the beauty, grandeur, and abundance of landscapes and other natural features."[20] This move by the Department of the Interior, now led by Trump appointee, Doug Burgum, aims to eliminate "wokeness" from parks, museums, and other federally managed sites. All of this undermines recent efforts made by collaborative teams of humanists and land managers to develop inclusive storytelling devices[21] and experiences on public lands. It is easy to feel overwhelmed by these threats and attempted erasures. However, we believe that this moment of proliferating crises also calls on us to gather, share, and preserve stories; build new archives; and *manage to love* (ourselves, each other, the world) through it all.

This collection seeks to join the "pockets of disruption"—to borrow a coinage from Joseph Gazing Wolf—that work to interrogate oppressive systems while also lovingly reconnecting humanity with the ecological communities to which they already belong.[22] Despite temptations to despair, we take heart in matriation movements that are returning Indigenous women to positions of land-based leadership. We take heart in reimagining ourselves, humans, as resourcing love, as rising to our potential as a keystone species. We listen with our heart to the words of Diné musician, scholar, and historian Lyla June:

> The Earth may be better off without certain systems we have created, but we are not those systems. We don't have to be, at least. What if I told you that the Earth needs us? What if I told you that we belong here?[23]

Notes

1 Deloria, Jr., *Custer Died for Your Sins*, 104.
2 Newland, *Federal Indian Boarding School Initiative Investigative Report*, vol. 1., 21.
3 Simpson, "Land as Pedagogy," 22–23.
4 Osuna and Tynan, *Storied Deserts*, 7.
5 Kimmerer, *The Serviceberry*, 22.

6 Gumbs, *Undrowned*, 8.
7 Britton, et al., "Wild Horse Roundups and Removals."
8 Britton, et al., "Wild Horse Roundups and Removals."
9 Britton, et al., "Wild Horse Roundups and Removals."
10 Tachine and Nicolazzo, *Weaving an Otherwise*, 7.
11 Fink, "Powerful Humanities Research Methods."
12 Robin, "Global Ideas in Local Places," 71.
13 Tachine and Nicolazzo, *Weaving an Otherwise*, 2.
14 Hooks, *All about Love*, 6.
15 brown, *Emergent Strategy*, 9.
16 Limón quoted in Poetry Society of America, "You Are Here."
17 Lyons, *X-Marks*, xii.
18 LaPier, "Land as Text," 1.
19 Kim, et al., "Enhancing Ecoliteracy," 2.
20 National Parks Conservation Association, "New Park Signs."
21 National Parks Foundation, "NPF Launches New Round."
22 Gazing Wolf, "Regrounding Democracy," 40:18–40:20.
23 June, "3,000-Year-Old Solutions to Modern Problems," 3:00–3:12.

References

Britton, Jennifer L., Abigail Del Grosso, Cassidy Ellis, and Christian Hunold. "Wild Horse Roundups and Removals: Affect, Gender, Interspecies Politics." *Humanimalia* 14, no. 1 (2023).

brown, adrienne marie. *Emergent Strategy: Shaping Change, Changing Worlds*. AK Press, 2017.

Deloria, Vine, Jr. *Custer Died for Your Sins: An Indian Manifesto*. Macmillan, 1969.

Fink, Lisa. "Powerful Humanities Research Methods Can Help Us Avoid Overburdening Tribes." *Park Science* 28, no. 1 (Summer 2024).

Gazing Wolf, Joseph. "Regrounding Democracy in Traditional Ecological Knowledge." *Green Dreamer Podcast*, Ep 433, September 3, 2024. https://www.greendreamer.com/podcast/joseph-gazing-wolf-heritage-lands-collective.

Gumbs, Alexis Pauline. *Undrowned: Black Feminist Lessons from Marine Mammals*. AK Press, 2020.

hooks, bell. *All about Love: New Visions*. William Morrow Paperbacks, 2018.

Johnson v. M'Intosh, 21 U.S. 543, Supreme Court of the United States, 1823.

June, Lyla. "3000-Year-Old Solutions to Modern Problems." *TED xKC*, 2022, Video, 13:28. https://www.youtube.com/watch?v=eH5zJxQETl4.

Lapier, Rosalyn. "Land as Text: Reading the Land." *Environmental History* 28, no. 1: 40–46. https://doi.org/10.1086/722618

Lyons, Scott Richard. *X-Marks: Native Signatures of Assent*. Minnesota University Press, 2010.

Kim, GaWoon, Rahul Teku Vaswani, Wanmo Kand, Miri Nam, and Dowon Lee. "Enhancing Ecoliteracy through Traditional Ecological Knowledge in Proverbs." *Sustainability* 9 (2017). https://doi.org/10.3390/su9071182.

Kimmerer, Robin Wall. *The Serviceberry: Abundance and Reciprocity in the Natural World*. Scribner, 2024.

National Parks Conservation Association. "New Park Signs Undermine Rangers, Aim to Erase History." *Press Release*, June 12, 2025.

National Parks Foundation. "NPF Launches New Round of Funding for Inclusive Storytelling Program." *National Parks Foundation Online*, May 8, 2024. https://www.nationalparks.org/news-and-updates/updates/npf-launches-new-round-funding-inclusive-storytelling-program.

Newland, Bryan. *Federal Indian Boarding School Initiative Investigative Report*, vol. 1. US Department of Interior, Indian Affairs, 2022.
Osuna, Celina, and Aidan Tynan. *Storied Deserts: Reimagining Global Arid Lands*. Routledge, 2024.
Poetry Society of America. "You Are Here: Poetry in Parks." *Poetry Society of America Online, 2025*. https://poetrysociety.org/programs/you-are-here.
Robin, Libby. "Global Ideas in Local Places: The Humanities in Environmental Management." *Environmental Humanities* 1 (2012): 69–84.
Simpson, Leanne Betasomasake. "Land as Pedagogy: Nishnaabeg Intelligence and Rebellious Transformation." *Decolonization: Indigeneity, Education & Society* 3, no. 3 (2014): 1–25.
Tachine, Amanda R., and Z. Nicolazzo. *Weaving Otherwise: In-Relations Methodological Practice*. Routledge, 2022.

Part 1

The Language of Management

1 Badlands Management

Joshua T. Anderson

The word "management" has precapitalist roots but has been co-opted by capitalist economics and sustained by settler-colonial land policies, leaving us with profit-first land relations that are evident from the grocery bill to the Farm Bill. But the linguistic roots of management take us back to horses and hands. The Italian *maneggiare*, meaning "to handle, touch, or control a horse," and Latin *manus*, or "hand," make phrases such as "hands-on management" a redundancy, while tracking the term's linguistic roots opens a range of possibilities about how to manage the badlands and prairies of the northern Great Plains.

In places like North Dakota, returning to the linguistic origins of management challenges us to restore our tactile relations with so-called "badlands" and to keep in touch with the plant and animal communities that endure in some of the baddest places in the more-than-human world. The Badlands of western North Dakota are a storied frontier that holds deep and ongoing histories of Indigenous peoples, including the Lakota, Cheyenne, Blackfeet, Crow, Arapaho, Anishinaabe, Ponca, and Omaha. The wild horses in Theodore Roosevelt National Park offer an unruly story of badlands management, igniting one of the toughest conservation debates in the American West. In April 2024, the NPS announced that wild horses would remain in the park after a years-long controversy spurred by an environmental assessment conducted by the NPS in 2021. But the story is far from over. "We still have no management plan," admitted Chris Kman of Chasing Horses Wild Horse Advocates, a group that has championed federal protections for wild horses, including the estimated 200 horses in the park's South Unit, many of which are believed to be descended from Sitting Bull's herd of Nokota.[1]

On the other side of the state, my home region in the Red River Valley offers another example of badlands management. The chipped-tooth border between North Dakota and Minnesota was once home to a diverse tallgrass prairie thrumming with bison, bear, wolves, and otters. It is the historic homelands of Anishinaabe and Dakota peoples who managed the prairie with controlled burns, seasonal hunting, and agricultural practices. Today, the region has been plowed, poisoned, and plundered by the extractive economies of sugar (sugarbeets), biofuel (corn and soybeans), and other low-quality cash crops.

DOI: 10.4324/9781003581185-3

In the winter of 2024, the Badlands Conservation Alliance collaborated with my office at the Walsh County Three Rivers Soil Conservation District (SCD) in northeastern North Dakota to develop a cross-state, experiential field camp for middle-school students that was set to be hosted in Teddy Roosevelt National Park in June of 2025. The field camp would have featured visits to working ranches and encounters with bison and wild horses, entangling students in thorny questions about how to manage North Dakota's threatened grasslands, wild horses, and working ranchlands. However, soon after reclaiming the White House, the Trump administration severely cut NPS staff and over $900 million from the NPS operational budget, leading us to cancel our trip to Teddy Roosevelt National Park.

As Dr. Surabhi Balachander writes elsewhere in this volume, land management in the United States is dependent upon our cultural relations with land and upon the administrative agencies tasked with managing and protecting lands, including the NPS, U.S. Fish and Wildlife Service (FWS), BLM, and U.S. Forest Service (USFS). The temperate grasslands of the Great Plains are the most threatened ecosystem on the continent. With the Trump administration's open hostility toward land management and regulatory agencies, it is crucial for us to foster grassroots efforts focused on prairie restoration at local and state levels. Specifically, we need committed land managers and land stewards capable of restoring prairie grasslands and conservationists and everyday citizens capable of re-storying our relationships to this vital ecosystem.

According to the FWS, an estimated 50 million acres of grasslands have been lost in the Great Plains to agricultural cultivation and non-native species (such as Kentucky bluegrass, smooth brome, and eastern red cedar) in the last decade alone.[2] The North Dakota Meadowlark Initiative formed in 2022 to address the catastrophic declines in grassland birds, with 53% of the population lost since 1970 (or approximately 720 million birds). North Dakota's state bird, the western meadowlark, has declined at 1.3% annually at the same time that US cattle ranchers have declined at a rate of 1% per year.[3] The fate of grassland birds is directly tied to the fate of ranchers, and collaborations between and among conservation organizations and ranchers have fostered a working lands approach to grassland conservation.

The working lands model builds partnerships between small and mid-sized ranchers and conservationists who work together to restore healthy grasslands. As Heather Husband, director of the ND Meadowlark Initiative, told me during an episode of my podcast series *Common Ground: A Prairie Podcast*: "what is good for the meadowlark is also good for working ranchlands." The meadowlark and other grassland birds rely upon functional, biodiverse prairies. And, as a disturbance dependent ecosystem, the prairie requires intensive, collaborative management by committed ranchers and other land stewards. It is a landscape reliant upon prescribed fire as much as rain—a landscape that co-evolved with bison, elk, and other grazing animals. Restoring our embattled prairie ecosystems requires rethinking our economic policies in the Great Plains through investments in grasslands management and grass-fed livestock. Rather than perpetuating the capitalist myth of perpetual growth, we must grow resilient working lands economies inspired by our tough, biodiverse prairie ecologies. Perhaps more than anything, this effort requires us to imagine and tell better stories. As Robin Wall Kimmerer contends in *Braiding Sweetgrass* (2013), the first step in ecological restoration is to re-story our relationships with land.

The work of restoring and re-storying prairie grasslands must take root at the local level. This need for local action has never been more evident than our current reality, wherein the Trump administration has enacted policies openly hostile toward agencies responsible for land management and environmental protections, cutting funds and staff from the NPS, FWS, USFS, and BLM, as well as the Environmental Protection Agency (EPA) and Natural Resources Conservation Service (NRCS). The SCD system that now employs me was born out of the losses incurred during the Dust Bowl, championing a model of farm and prairie management rooted in local governance and grassroots action. In *A Sand County Almanac* (1949), Aldo Leopold wrote that soil districts were "a beautiful piece of social machinery." But he warned that we were only fulfilling half of our original obligation, explaining that SCDs were not merely established to provide cost-share funding, education and outreach, and on-the-ground support. Crucially, SCDs were also implemented to establish local rules for land-use that would be enforceable by municipal and county laws.

As Leopold observed:

> When one asks why no rules have been written, one is told that the community is not yet ready to support them; education must precede rules. But the education actually in progress makes no mention of obligations to land over and above those dictated by self-interest. The net result is that we have more education but less soil, fewer healthy woods, and as many floods as in 1937.[4]

In order to nurture a prairie-based economy, we must find common ground through ongoing collaboration and challenge capitalist modes of thinking about our relations to land through action and effective policymaking. The prairie is perhaps our best example of a literal common ground: a landscape that is home to a vast range of human and more-than-human communities. Often referred to as "flyover country" by those who never set foot here, the Great Plains is truly flyover country for migratory birds, pollinators, and waterfowl. The prairies of the northern Great Plains have deep roots and a wide range, both culturally and ecologically, and restoring prairies is perhaps our best hope against the threats posed by climate change and species extinction.

Inspired by the many teachings of the tallgrass prairie, my office recently partnered with Canadian-American artist Austen Camille to host experiential art workshops titled "Thinking Like a Prairie." Funded by a grant from the North Dakota Natural Resources Trust, our project took high school students to a local bison ranch and to a restored prairie site, where we collected grasses and wildflowers that we turned into prairie paper and soils we converted into paint pigments. In an article about the "Thinking Like a Prairie" project by journalist Michael Standaert, students offered firsthand accounts about their experiences. As student John Calles observed: "I knew prairies existed, but I didn't really think much about them. . . . We walked through the prairie picking out plants for the paper we made and walking through it I understood how thick and lush it was. It was really cool." Similarly, student Bianca Ramos noted how the project taught her to notice the abundance of life in the prairie, where "there are a lot of different plants, and they're homes and food for insects and animals."[5]

By recognizing the prairie as a teacher, we are working to challenge Western educational paradigms that separate students from the lands and waters where we live and to cultivate new generations of land managers prepared to think, dream, and act in ways grounded in lessons learned from deep ecology. With prairie as a classroom, students in my home county have begun to learn the science and the stories that can help us to re-story our current relations with land. And, as we continue these efforts, we hope to build reciprocal relations with Indigenous Nations, including the nearby Spirit Lake Nation of Dakota and the Turtle Mountain Band of Chippewa. In so doing, we understand that land management is best when it is cross-cultural and multigenerational, fostering multispecies relations and affording lifelong lessons from and with the land.

Given the central importance of stories, it is vital for us to take a hard look at the language we use to name our relations to place. Language is intimately tied to law, and if we are to challenge the current laws of the land, we must understand and, at times, resist how we express our relations to land through spoken and written words. The "Thinking Like a Prairie" project is just one example of our efforts to offer students critical lessons in this form of ecological literacy, where we learn how to read the land and water from perspectives beyond the human. As student John Calles recalled in his interview with Standaert: "I learned there's this grass called Big Bluestem, and it uses fire to germinate."[6] Similarly, many students reported that listening to the restored prairie site, thrumming with insects and birdsong, had a profound effect on them, especially when they compared it to the silence of the neighboring sugar beet fields.

By paying closer attention to the way human and more-than-human languages shape our relations with land, we can become better equipped to reimagine and reshape the future of land management. But, to restore and manage an ecoregion as diverse as the prairie, we will also need to become better students of ecological history.

Take this concluding story as an example.

While students dug into prairie soils, preparing to render the clay into paint pigments, I asked them to stop for a moment and imagine the Ice Age. I asked them to consider how our home region was bulldozed flat by glaciers and to feel the remnants of the glacial lakebed that covered their soiled hands. As students imagined in geological time, we listened to the flute-like songs of a meadowlark and the hum of native bumblebees. We noted the seasonal change from late summer to early fall in the bloom of goldenrod and the sonorous honk of migratory geese arrowing south for the winter.

"Have any of you heard of Alexander Henry?" I asked.

A student blurted out "who's that?"

"Okay," I said. "Let me tell you a story."

"Imagine this land before it was called Walsh County," I began. "Before it was North Dakota. Imagine we are standing here during the time of the Lewis and Clark expedition. That's when Alexander Henry was here, camping along with his men right here, by the Red River."

I pulled up a passage on my phone from Henry's journals, dated September 18, 1800, when he climbed an oak tree and described what he saw:

> I took my usual morning view from the top of my oak and saw more buffaloes than ever. They formed one body . . . as far as the eye could reach. They were moving southward slowly and the meadow seemed as if in motion.[7]

I told the students that the "Three Rivers" in the Walsh County Three Rivers SCD refers to the three branches of the Park River, which, Henry explained, "derives its name from the fact that the Assiniboines once made a park or [canal] pound on this river for buffalo."[8] I explained how, today, my office is just down the road from the Alexander House, the roadside motel and liquor store that pays tribute to the nineteenth-century explorer, complete with a sign portraying Alexander Henry in a coonskin cap, a rifle cradled in the crook of his arm. The words "Alexander House" are written in a font that resembles cut logs.

As students listened, it was impossible to ignore the obvious and dramatic changes all around us. The once endless grassland Henry described has been nearly lost to a history of exploitation and violence. Gone are the enormous herds of roaming bison. The bears are overhunted for their fur and fat. The stands of tallgrass reduced to the ditches and dwindling cattle pastures.

However, I pointed out the big bluestem (*Andropogon gerardii*) still holding tough in the ditches, asking students to look closely at its distinctive purplish tint and turkey track seedheads (or rames). "It's beautiful," one student remarked. "Wild," another student noted, as she touched the seedheads, collecting seeds for our prairie paper-making workshop.

During Austen Camille's painting lesson later that week, she asked students to paint big bluestem and prairie liatris on the paper they'd just made with prairie grasses and flowers. The paints rendered from soils formed by an Ice Age and thousands of years of tallgrass prairie.

"How does it feel to paint with materials that are over 10,000 years old?" Austen asked.

"It's pretty crazy," a student said, with a big grin.

Getting our hands dirty with these long-lived prairie plants and ancient soils can inspire us to take a deeper view of our conservation and grassland management efforts. To ask: how can the living roots of history inspire us to change farm policies and practices?

Similarly, land management is often a dirty job, but the linguistic origins of the word "management" remind us that the power to change our story is in our hands. This is true even in a landscape now dominated by erosion and white knuckles. Even where histories of exploitation and violence appear too deep to uproot. Where the hopeful future seems too distant on the prairie's horizon.

Notes

1 Beach, "Wild Horses."
2 United States Fish and Wildlife Service, "Central Grasslands."
3 Meadowlark Initiative, "What Is."
4 Leopold, *A Sand County Almanac*, 246.
5 Standaert, "Students 'Think Like a Prairie.'"
6 Standaert, "Students 'Think Like a Prairie.'"
7 Henry, *The Manuscript Journals of Alexander Henry*, 99.
8 Henry, *The Manuscript Journals of Alexander Henry*, 93.

Works Cited

Beach, Jeff. "Wild Horses to Stay in National Park; Management Plan Needed, Advocate Says." *North Dakota Monitor*, April 25, 2024. https://northdakotamonitor.com/briefs/wild-horses-to-stay-in-national-park-management-plan-needed-advocate-says/.

Henry, Alexander. *The Manuscript Journals of Alexander Henry, Fur Trader of the Northwest Company, and of David Thompson, Official Geographer and Explorer of the Same Company 1799–1814 Exploration and Adventure among the Indians on the Red, Saskatchewan, Missouri, and Columbia Rivers. In Three Volumes. Vol. I. The Red River of the North*. Francis P. Harper, 1897. https://open.library.ubc.ca/collections/bcbooks/items/1.0363301.

Leopold, Aldo. *A Sand County Almanac: With Essays on Conservation from Round River*. Oxford University Press, 1949.

Meadowlark Initiative. "What Is the Meadowlark Initiative?" *North Dakota Game and Fish Online*. Accessed March 18, 2026. https://gf.nd.gov/meadowlark-initiative.

Standaert, Michael. "Students 'Think Like a Prairie' with Hands-On Art Experience." *Grand Forks Herald*, September 16, 2025.

United States Fish and Wildlife Service. "Central Grasslands Conservation—About Us." FWS.gov. N.d. https://www.fws.gov/program/central-grasslands-conservation/about-us.

2 Relationality

Lisa Fink and Lydia Cagluaq Agnus

Embedded in Indigenous names are cosmological beliefs that illustrate Indigenous relationality. For example, Yup'ik names provide a deeper connection to people, place, culture, and the environment. The word "Yup'ik" means "real person" or "genuine person," combining *yuk* (person) and *pik* (real or genuine). Yup'ik culture places a strong emphasis on community, reciprocity, and respect for all living beings. Yup'ik names feed the soul and represent Yup'ik identity and relationships. As Potawatomi botanist Robin Wall Kimmerer explains, "Each name calls up a story and connects us to the past and present of [a] place every time we say it."[1] When settlers renamed Qaluyaar "Nelson Island," it erased many of the stories and beliefs of Yup'ik culture relational to our environment. In what follows, we briefly outline *relationality* as scholarly disciplines have addressed it, offer two approaches to relationality from the Yupiit and Anishinaabeg, discuss colonial and capitalist relations that have disrupted other forms of relationality, and analyze some North American literature that imagines alternative relationalities for more just futures.[2] As a Yup'ik scholar and a settler scholar both based in Alaska, we engage with the keyword *relationality* as "related comrades."[3]

Recent work in Indigenous studies has sought to theorize relationality—the ways humans, nonhumans, the land, and all things are interconnected—and discuss ways for scholars to engage in forms of relationality that exceed capitalist relations. Matt Wildcat (Ermineskin Cree) and Daniel Voth (Métis) explain that Indigenous Nations and collectivities cultivate diverse approaches to relationality that include shared understandings across the global Indigenous world, context-specific concepts from particular Native Nations, and inter-Indigenous relationships.[4] These approaches include relations both among humans and between humans and the more-than-human world. Moreover, relationality is tied to values that shape a community's economic and political interactions.[5] It is also connected to knowledge production. For example, Leanne Betasamosake Simpson (Michi Saagiig Nishnaabeg) explains that knowledge generation occurs through "relational reciprocity" with *Aki*, meaning land.[6] Additionally, relationality is often expressed through the concept of "kinship," which Kyle Powys Whyte (Potawatomi) describes as "an ethic of shared responsibility."[7] Indigenous notions of kinship extend beyond

DOI: 10.4324/9781003581185-4

the heteropatriarchal nuclear family to one's extended family, community, plants, animals, land, and water, as well as past and future generations. Exploring relationality through lessons from Country, Lauren Tynan (Pairrebenne Trawlwoolway) adds that relationality is "a living practice" that includes "responsibilities with kin."[8] More recently, scholars like Jodi Byrd (Chickasaw), Billy-Ray Belcourt (Driftpile Cree), and Kim TallBear (Sisseton Wahpeton Oyate) have sought to explore queer and trans relationalities, bridging Indigenous and queer studies.[9]

Recent decades have seen a relational turn in other disciplines across the humanities, social sciences, and natural sciences. For example, notions of relationality emerging in feminist science and animal studies consider encounters with companion species.[10] Scholars of sustainability science recognize Indigenous knowledge as a source of insight, arguing for crediting the Indigenous thinking and practices that have embodied "deep relationality" since time immemorial.[11] Despite the prevalence of the term across disciplines, many dominant research practices continue to be extractive.[12] As Jodi Byrd notes, the shared language of relationality emerging as a key concept in a range of fields can "giv[e] the illusion that there is a conversation underway when, in fact, the shared vocabulary is a structural violence or outright disavowal after all."[13] For Native and non-native scholars, methodologies can genuinely integrate relationality by not only prioritizing relationships but also critically analyzing "how we create relationships that . . . have the power to further Indigenous self-determination."[14] Aileen Moreton-Robinson (Goenpul) observes that Indigenous social research paradigms, which are inherently based on relationality, offer "new approaches to collaborative and participatory research by seeking to make visible our connectedness with earth and with each other."[15] Yet, because "the discourse of relationship can be invoked to mask or legitimate [sic] oppressive practices," Gina Starblanket (Cree/Saulteaux) and Heidi Kiiwetinepinesiik Stark (Turtle Mountain Ojibwe) suggest applying a relational lens to scholarship through four critical concepts—knowledge, gender, land, and modernity—that target colonizing logics and pursue "relationships as generative, active processes."[16] Next, we explore Yup'ik and Anishinaabe examples of relationality as it applies to land management.

In the Yup'ik language, *Ella* means weather, awareness, and world, and *Ellam yua* means the spirit or person of the universe who is everywhere.[17] For Yup'ik people, the land possesses its own spirit because everything has a spirit. Everything is respected because of *Ellam yua*. Including *Ellam yua* in this discourse recognizes the respect and responsibility that Yup'ik people have for the spirits of animals, plants, land, and waterways. It is said that "a Yup'ik person comes into the world as a part of a complex web of kinship, including living and dead humans and animals."[18] The Yup'ik worldview is to live in harmony and reciprocity with the natural environment because everything has a spirit.

In Yup'ik culture, names are deeply tied to reincarnation, which plays a significant role in Yup'ik naming traditions. The spirit of the person—and the essence of what it means to be human—passes through names. Thus, personhood is created through the cosmological belief of reincarnation.[19] Yup'ik naming traditions are tied to the relationships created when one is named after someone who recently passed away, allowing the deceased to live through the newly named infant. The belief in reincarnation has a significant association with sustainable land management through its connections with living, nonliving, natural, and spiritual worlds. This cosmological approach informs the Yup'ik worldview of caring for the environment. This belief shapes the values and actions toward the treatment of the land and its inhabitants, ensuring that future generations can continue to live in harmony with the land. Naming a child with a Yup'ik name reinforces the idea of the cycle of life after death that connects the past, present, and future. This belief in reincarnation promotes long-term sustainable practices for generations because the Yup'ik return to this earth. When a child is named after an ancestor, they not only honor that person; they also honor all people who carried that name. The name continues their presence in the world. This reinforces Yup'ik relationality, especially kinship relationships, because we are always in relation with our ancestors, community, environment, and the spiritual world. A name becomes a living thread in a complex web of relationships and responsibility.

By continuing naming traditions, Yup'ik people preserve not only the individual and communal identity of what it means to be Yup'ik but also the values and cultural knowledge of land stewardship. As anthropologist Ann Fienup-Riordan explains,

> Yup'ik ontological and epistemological principles of interspecies responsibility regularly conflict with the human-centric, dualistic western paradigm. Many Yup'ik people still understand their environment as a world occupied by persons, where compassion and restraint daily characterize all interpersonal interaction.[20]

In other words, bountiful harvests from land and waterways create relationships and reciprocity. Here, since time immemorial, my (Lydia's) ancestors built a long-standing relationship with our land, lakes, rivers, plants, and animals.

Anishinaabe forms of relationality are both similar to and distinct from Yup'ik forms, in part because they are geographically situated. The Anishinaabeg are the peoples known as the Three Fires Confederacy—the Ojibwe, Potawatomi, and Odawa—who currently live across a vast region from the Great Lakes to the Plains in the present-day United States and Canada, as well as in other urban and rural communities throughout North America. In *Ojibway Heritage*, elder and linguist Basil H. Johnston (Anishinaabe) explains that Gichi Manidoo created the world in a particular order: first, the sun, moon, earth, and stars; next, the plant world; and, third, the animal beings.[21] Only later did spirit-woman (or Sky Woman) arrive, and then the first human beings were born and grew up, thanks to the assistance the animals provided.[22] To the Anishinaabeg, plant beings are "elder sisters," and animal beings are "elder brothers."[23] As this order suggests, plants "could exist alone; they were not dependent upon other beings for their existence or well-being."[24] On the other hand, humans could not survive without plants and animals. Animals have knowledge that humans do not, such as the ability to sense changes in the environment and seasons. Humans must rely on animals for this knowledge.[25] This Anishinaabe creation story engenders humility toward everything else in Creation because the plants and animals are older and have more knowledge than humans.[26] Importantly, this story exemplifies a key aspect of Anishinaabe ontologies: kinship between humans and the more-than-human world. Alongside kinship, the Anishinaabeg recognize interdependence with all beings as a critical feature of their social relations.

Anishinaabe relationality also emphasizes the importance of naming as a relational practice. In another creation story, Gichi Manidoo breathed life into the Original Man, a being who is part man and part spirit named Nanabozho.[27] Nanabozho was created last of all beings, and Gichi Manidoo gave him several tasks, his "Original Instructions":[28] to walk through the world so "that each step is a greeting to Mother Earth," to learn from the world how to be human, and to learn the names of all beings by carefully watching how they live and speaking with them to learn about their gifts.[29] Residing in Indigenous ecological knowledges, Original Instructions teach the Anishinaabeg "how to live on the earth in a good way" with other living and non-living beings.[30] Original Instructions also lie in the natural world where "the real intelligence dwells," reminding people to practice humility, interconnection, and interdependence.[31] These stories illustrate the Anishinaabeg's orientation to the world and key tenets of humility, curiosity, and the "relational responsibilities" of learning the names of everything that came before them to understand their role within the wider circle of relationality.[32] Moreover, names in Ojibwemowin[33] often indicate relationships between humans and the more-than-human world. For example, the black ash tree (*Fraxinus nigra*) is *baapaagimaak*, which translates in English to "the tree that you pound" to get splints for weaving baskets and materials for making *aagim* or snowshoes. The name shows the Ojibwe relationship to this tree, highlighting that language revitalization is closely tied to self-determination in land management. In addition, naming indicates a relationship tied to specific practices of respect.

Capitalism and colonialism disrupt Indigenous approaches like the ones we've described above, with critical implications for Indigenous peoples and their lifeways. Land developers and governing agencies frequently access and interrupt Indigenous lands and resources for economic development. As Glen Coulthard (Yellowknives Dene) argues, what Marx theorized as the historical process of "primitive accumulation" is actually an ongoing practice of dispossession that disrupts relationalities held in the land—relationalities that inform Indigenous collectivities' economic and political orders.[34] Indigenous resistance and resurgence work "to reconnect Indigenous bodies to land through the practices and forms of knowledge that these practices continually generate."[35] In contrast, settler-racial capitalist relations with land emphasize individualism, independence, personal gain, and human supremacy. These practices break (or preclude the conditions of possibility for) other forms of relation, especially those that understand relations as a vast network of interconnected relationalities between lands, waters, skies, humans, plants, animals, ancestors, and more. In breaking those relations, processes of settler-racial capitalism—especially those entangled with land management—serve to ensure settler futurity at the expense of Indigenous and Black Life.

Moreover, primitive accumulation includes changing the relational meaning of the land. For example, Kimmerer discusses the ways settlers have sought to supplant Indigenous heterogenous relationalities. She contrasts Potawatomi approaches that understand "Land as Teacher/Land as Healer" with settler-racial capitalist relations of "Land as Property"—that is, land as a commodity for exchange.[36] Similarly, Simpson defines "land as pedagogy," a place that we learn from, with, and through.[37] For the Anishinaabe, "land, Aki, is both context and process," and this "context is the curriculum."[38] From the Alaska Native perspective, land is not for sale; rather, people need the land to survive. For example, Yup'ik people view *nuna,* land, not merely as a resource but as a spiritual being with awareness. *Nuna* belongs to itself. Conversely, instead of relating to Land as a place of healing and learning, colonization relates to land as a place of extraction, control, and domination through processes of commodification and data culture, to name two.[39] Max Liboiron (Michif-settler) conveys this difference in relationality by drawing a distinction between small-l land, referring to "the concept from a colonial worldview, whereby landscapes are common, universal, and everywhere, even with great variation," and uppercase-l Land, indicating "a proper name indicating a primary relationship."[40] Additionally, as Allison Weir explains, by instituting heteropatriarchal and racialized forms of property ownership, colonization obscures and undermines these alternative forms of relationality through "colonial unknowing." This process undermines the power of Indigenous women and Two Spirit people, thereby consolidating power in the hands of white, male, cis-het settlers.[41]

Understanding "Land as Property" cannot be divorced from the ways whiteness has become a form of property protected by US law within colonialist and capitalist forms of relationality. Legal scholar Cheryl I. Harris outlines the rights of whiteness as property under US law, including the right to status and property, such as the elevated status of White people under US white supremacy and the right to exclude BIPOC groups from legal protection.[42] Western-colonial land management regimes uphold the right of whiteness to use public land for its benefit. For example, national parks were created as spaces of escape and leisure for white elites by dispossessing local Indigenous people and poor and working-class people who were part of the resident subsistence economy.[43] In a recent example, the US Army Corps of Engineers

(USACE) issued a permit for the Dakota Access Pipeline (a name itself indicating a resource-based relationship with land) in opposition to the Standing Rock Sioux Tribe.[44] That the USACE has the elevated status to write (and fail to disclose) the Environmental Impact Statement and grant permits around water crossings without consent from the sovereign nation of the Standing Rock Sioux highlights how whiteness-as-property fortifies layers of power and exclusion under current US land management regimes. These regimes also exclude Indigenous people from accessing and managing their land. For instance, Alaska Native people have experienced the deleterious effects of Western land management practices that limited their use of and marginalized their connection to the land. The Alaska Native Claims Settlement Act (ANCSA) of 1971 extinguished aboriginal land rights and shifted the ownership and management of traditional Alaska Native systems to a corporate model. Western land management follows the model of land as property and, relatedly, the rights of White settlers to enjoy and use that property and pass those rights on to their descendants, ensuring their continued ownership.

As this collection demonstrates, North American literature can counter colonial unknowing and illustrate alternative relationalities critical to more effective and just land management that must include Native Nations managing their lands. Novels and poetry by writers such as Lily H. Tuzroyluke (Iñupiaq), Sheila Watt-Cloutier (Inuit), Octavia Butler, Ofelia Zepeda (Tohono O'odham), Leslie Marmon Silko (Laguna Pueblo), and Craig Santos Perez (CHamoru), to name a few among many, represent and imagine relationalities that can provide models for creating more just futures.[45] Scholars like Tiffany Lethabo King, Juliana Hu Pegues, Sarah D. Wald, and others reveal relationalities beyond the settler-colonial–capitalist (read: white, cisheteropatriarchal) relations that dominate North American land management.[46] Dominant systems of management prioritize capitalist relations of extraction and commodification while narratives of settlers' local, place-based relationships naturalize "fantasies of colonial belonging."[47] Despite this, some federal agencies, including the US NPS where we work and with whom we partner, have advanced efforts to collaborate with Native Nations, as per their federal trust responsibility.[48] Literature also helps to vivify traditional knowledge approaches—place-based, embodied, multigenerational knowledge that federal land managers too often fail to consider. Western land managers must understand Indigenous relationalities to better collaborate with communities who have carried forward such land-based knowledge for millennia. They must seek to work alongside Indigenous approaches when developing agency policies and include Indigenous communities across Turtle Island (and all of Abiayala[49]) in decision-making and development of community-driven initiatives.

Notes

1 Kimmerer, *Gathering Moss*, 2–3.
2 "Yupiit" and "Anishinaabeg" are plural forms of "Yup'ik" and "Anishinaabe," respectively.

3 Coulthard and Simpson, "Grounded Normativity/Place-Based Solidarity," 250.
4 Wildcat and Voth, "Indigenous Relationality," 476.
5 Wildcat and Voth, "Indigenous Relationality," 477.
6 Simpson, *As We Have Always Done*. See also Ziavra's chapter in this volume.
7 Whyte, "Time as Kinship," 40.
8 Tynan, "What Is Relationality?," 601–604.
9 Byrd, "What's Normative Got to Do with It?"; Belcourt, "Can the Other of Native Studies Speak?"; TallBear, "Making Love and Relations Beyond Settler Sex and Family."
10 Haraway, *The Companion Species Manifesto*.
11 Gould, Martinez (Mescalero Apache–Pennsylvania Dutch), and Hoelting, "Exploring Indigenous Relationality to Inform the Relational Turn in Sustainability Science."
12 Tynan, "What Is Relationality?," 604–606.
13 Byrd, "What's Normative Got to Do with It?," 106.
14 Wildcat and Voth, "Indigenous Relationality," 480.
15 Moreton-Robinson, "Relationality," 75.
16 Stark and Starblanket, "Towards a Relational Paradigm," 175.
17 Ayuncrak (Yup'ik), et al., "Yup'ik Culture and Context in Southwest Alaska."
18 Litecky, "The Dwellers Between," 33.
19 Fienup-Riordan, *Hunting Tradition in a Changing World*.
20 Fienup-Riordan and Rearden (Yup'ik), *Ellavut*, 38–39.
21 Johnston, *Ojibway Heritage*, 33–47. Following Simpson's example, I (Lisa) provide brief sketches of these creation stories based on printings by elders. It is not ethically appropriate for the stories to be shared out of context. Simpson, *Dancing on Our Turtle's Back*, 35.
22 Johnston, *Ojibway Heritage*, 49–50.
23 Johnston, *Ojibway Heritage*, 58; Simpson, "Our Elder Brothers," 73–87.
24 Simpson, "Our Elder Brothers," 33.
25 Johnston, *Ojibway Heritage*, 52–53.
26 See also Watts (Haudenosaunee and Anishinaabe), "Indigenous Place-Thought and Agency amongst Humans and Non-Humans."
27 Kimmerer, *Braiding Sweetgrass*, 205; Simpson, *Dancing*, 39.
28 Nelson (Turtle Mountain Ojibwe), ed., *Original Instructions*.
29 Kimmerer, *Braiding Sweetgrass*, 206–208.
30 Ausubel, "Preface: Remembering the Original Instructions," xxi.
31 Ausubel, "Preface: Remembering the Original Instructions," xxii.
32 Some Anishinaabeg use the term "relational responsibilities" when referring to "land management" or "land stewardship" to more accurately represent the practice of Anishinaabe relations with the more-than-human world. Interview by Lisa Fink with band member/harvester/educator, 21 September 2021. Since 2021, I (Lisa) have conducted community-driven research, in partnership with the Great Lakes Indian and Wildlife Commission, with nine Anishinaabe nations using decolonizing and anticolonial research methods as directed by my research partners.
33 Ojibwemowin (or Anishinaabemowin) is the language the Ojibwe speak.
34 Coulthard and Simpson, "Grounded Normativity/Place-Based Solidarity," 251. Karl Marx's theory of "primitive accumulation" in *Capital* vol. 1 identifies the violent transformation of noncapitalist forms of life, such as Indigenous societies, peasants, and other small-scale, self-sufficient agricultural producers, into capitalist social relations, which ripped from them their means of production and subsistence. In *Red Skin, White Masks,* Coulthard emphasizes that this process includes ripping noncapitalist societies from the land as the source of their livelihood (7).
35 Coulthard and Simpson, "Grounded Normativity/Place-Based Solidarity," 254.

36 Kimmerer, *Braiding Sweetgrass*, 329–335.
37 Simpson, *As We Have Always Done*, 145–174.
38 Simpson, *As We Have Always Done*, 151, 155.
39 Grossman, *Immeasurable Weather*.
40 Liboiron, *Pollution Is Colonialism*, 7n19.
41 Weir, *Decolonizing Freedom*, 155. For example, Federal Indian law and policies sought to remake Indigenous kinship structures, genders, and sexualities into what was recognizable for heteronormative Anglo-Americans. Rifkin, *When Did Indians Become Straight?*
42 Harris, "Whiteness as Property."
43 Spence, *Dispossessing the Wilderness*; Jacoby, *Crimes Against Nature*.
44 Estes (Kul Wicasa), *Our History is the Future*.
45 Tuzroyluke, *Sivulliq: Ancestor*; Sheila Watt-Cloutier, *The Right to Be Cold*; Butler, *Parable of the Sower*; Ofelia Zepeda, *Ocean Power*; Silko, *Ceremony*; Perez, *from unincorporated territory [åmot]*. See also Vargas's, Brown's, and Carpenter's essays in this volume.
46 King, *The Black Shoals*; Hu Pegues, *Space-Time Colonialism*; Wald, *The Nature of California*.
47 Hooley, *Against Extraction*, 25.
48 Hill, et al., eds., *National Parks, Native Sovereignty*.
49 *Abiayala* is an original and ancestral name of the Western Hemisphere. It means "Land of Plenitude and Maturity" in the language of the Guna, an Indigenous Nation in the region of Guna Yala, formally known as San Blas in present-day Panama. It is also a philosophical and political concept for reimagining a transhemispheric Indigeneity by reclaiming Indigenous philosophies and knowledges to advance collective resurgence and liberation. See Keme and Coon, "For Abiayala to Live, the Americas Must Die."

Bibliography

Ausubel, Kenny. "Preface: Remembering the Original Instructions." In *Original Instructions: Indigenous Teachings for a Sustainable Future*, edited by Melissa K. Nelson. Bear and Company, 2008.

Ayunerak, Paula, Deborah Alstrom, Charles Moses, James Charlie, and Stacy M. Rasmus. "Yup'ik Culture and Context in Southwest Alaska: Community Member Perspectives of Tradition, Social Change, and Prevention." *American Journal of Community Psychology* 54 (2014): 91–99. https://doi.org/10.1007/s10464-014-9652-4.

Belcourt, Billy-Ray. "Can the Other of Native Studies Speak?" *Decolonization: Indigeneity, Education, and Society* (blog), February 1, 2016. https://decolonization.wordpress.com/2016/02/01/can-the-other-of-native-studies-speak/.

Butler, Octavia E. *Parable of the Sower*. Four Walls Eight Windows, 1993.

Byrd, Jodi A. "What's Normative Got to Do with It? Toward Indigenous Queer Relationality." *Social Text* 38, no. 4 (2020): 105–123. https://doi.org/10.1215/01642472-8680466.

Coulthard, Glen. *Red Skin, White Masks: Rejecting the Colonial Politics of Recognition*. University of Minnesota Press, 2014.

Coulthard, Glen, and Leanne Betasamosake Simpson. "Grounded Normativity/Place-Based Solidarity." *American Quarterly* 68, no. 2 (2016): 249–255. https://doi.org/10.1353/aq.2016.0038.

Estes, Nick. *Our History Is the Future: Standing Rock versus the Dakota Access Pipeline, and the Long Tradition of Indigenous Resistance*. Haymarket Books, 2024.

Fienup-Riordan, Ann. *Hunting Tradition in a Changing World: Yup'ik Lives in Alaska Today*. Rutgers University Press, 2000.
Fienup-Riordan, Ann, and Alice Rearden. *Ellavut: Our Yup'ik World and Weather: Continuity and Change on the Bering Sea Coast*. University of Washington Press, 2013.
Gould, Rachelle K., Doreen E. Martinez, and Kristin R. Hoelting. "Exploring Indigenous Relationality to Inform the Relational Turn in Sustainability Science." *Ecosystems and People* 19, no. 1 (2023): 2229452. https://doi.org/10.1080/26395916.2023.2229452.
Grossman, Sara J. *Immeasurable Weather: Meteorological Data and Settler Colonialism from 1820 to Hurricane Sandy*. Duke University Press, 2023.
Haraway, Donna. *The Companion Species Manifesto: Dogs, People, and Significant Otherness*. Prickly Paradigm Press, 2003.
Harris, Cheryl I. "Whiteness as Property." In *Critical Race Theory: The Key Writings that Formed the Movement*, edited by Kimberlé Crenshaw, Neil Gotanda, Gary Peller, and Kendall Thomas. New Press, 1995.
Hill, Christina Gish, Matthew J. Hill, and Brooke Nelly, eds. *National Parks, Native Sovereignty: Experiments in Collaboration*. University of Oklahoma Press, 2024.
Hooley, Matt. *Against Extraction: Indigenous Modernism in the Twin Cities*. Duke University Press, 2024.
Hu Pegues, Juliana. *Space-Time Colonialism: Alaska's Indigenous and Asian Entanglements*. UNC Press Books, 2021.
Jacoby, Karl. *Crimes against Nature: Squatters, Poachers, Thieves, and the Hidden History of American Conservation*. University of California Press, 2014.
Johnston. *Ojibway Heritage*. University of Nebraska Press, 1999.
Keme, Emil, and Adam Coon. "For Abiayala to Live, the Americas Must Die: Toward a Transhemispheric Indigeneity." *Native American and Indigenous Studies* 5, no. 1 (2018): 42–68. https://dx.doi.org/10.1353/nai.2018.a704738.
Kimmerer, Robin Wall. *Braiding Sweetgrass: Indigenous Wisdom, Scientific Knowledge and the Teachings of Plants*. Milkweed Editions, 2013.
Kimmerer, Robin Wall. *Gathering Moss: A Natural and Cultural History of Mosses*. OSU Press, 2003.
King, Tiffany Lethabo. *The Black Shoals: Offshore Formations of Black and Native Studies*. Duke University Press, 2019.
Liboiron, Max. *Pollution is Colonialism*. Duke University Press, 2021.
Litecky, Ahnie. "The Dwellers Between: Yup'ik Shamans and Cultural Change in Western Alaska." MA Thesis, University of Montana, 2011. https://scholarworks.umt.edu/etd/982.
Marx, Karl. *Capital*, vol. 1. Penguin, 1990.
Moreton Robinson, Aileen. "Relationality." In *Sources and Methods in Indigenous Studies*, edited by Chris Andersen and Jean M. O'Brien, 69–77. Routledge, 2016.
Nelson, Melissa K., ed. *Original Instructions: Indigenous Teachings for a Sustainable Future*. Bear & Company, 2008.
Perez, Craig Santos. *from unincorporated territory [åmot]*. Omnidawn Publishing, 2023.
Rifkin, Mark. *When Did Indians Become Straight? Kinship, the History of Sexuality, and Native Sovereignty*. Oxford University Press, 2011.
Silko, Leslie Marmon. *Ceremony*. Viking Press, 1977.
Simpson, Leanne Betasamosake. *As We Have Always Done: Indigenous Freedom through Radical Resistance*. University of Minnesota Press, 2017.
Simpson, Leanne Betasamosake. *Dancing on Our Turtle's Back: Stories of Nishnaabeg Re-Creation, Resurgence, and a New Emergence*. Arbeiter Ring Publications, 2011.

Simpson, Leanne Betasamosake. "Our Elder Brothers: The Lifeblood of Resurgence." In *Lighting the Eighth Fire: The Liberation, Resurgence, and Protection of Indigenous Nations*. AK Press, 2008.

Spence, Mark David. *Dispossessing the Wilderness: Indian Removal and the Making of the National Parks*. Oxford University Press, 1999.

Stark, Heidi Kiiwetinepinesiik, and Gina Starblanket. "Towards a Relational Paradigm: Four Points for Consideration: Knowledge, Gender, Land, and Modernity." In *Resurgence and Reconciliation: Indigenous-Settler Relations and Earth Teachings*, edited by Michael Asch, John Borrows, and James Tully, 175–207. University of Toronto Press, 2018.

TallBear, Kim. "Making Love and Relations Beyond Settler Sex and Family." In *Making Kin Not Population*, edited by Adele Clarke and Donna Haraway, 145–164. Prickly Paradigm Press, 2018.

Tuzroyluke, Lily H. *Sivulliq: Ancestor*. Epicenter Press, 2023.

Tynan, Lauren. "What Is Relationality? Indigenous Knowledges, Practices and Responsibilities with Kin." *Cultural Geographies* 28, no. 4 (2021): 597–610. https://doi.org/10.1177/14744740211029287.

Wald, Sarah D. *The Nature of California: Race, Citizenship, and Farming since the Dust Bowl*. University of Washington Press, 2016.

Watt-Cloutier, Sheila. *The Right to Be Cold: One Woman's Fight to Protect the Arctic and Save the Planet from Climate Change*. University of Minnesota Press, 2018.

Watts, Vanessa. "Indigenous Place-Thought and Agency Amongst Humans and Non-Humans (First Woman and Sky Woman Go on a European World Tour!)." *Decolonization: Indigeneity, Education, and Society* 2, no. 1 (2013): 20–34.

Weir, Allison. *Decolonizing Freedom: Studies in Feminist Philosophy*. Oxford University Press, 2024.

Whyte, Kyle Powys. "Time as Kinship." In *The Cambridge Companion to Environmental Humanities*, edited by Jeffrey Cohen and Stephanie Foote, 39–55. Cambridge University Press, 2021.

Wildcat, Matt, and Daniel Voth. "Indigenous Relationality: Definitions and Methods." *AlterNative: An International Journal of Indigenous Peoples* 19, no. 2 (2023): 475–483. https://doi.org/10.1177/11771801231168380.

Zepeda, Ofelia. *Ocean Power*. University of Arizona Press, 1995.

3 Reclamation

Quinn Grover

In 2008, the United States Bureau of Reclamation (USBR) published a series of essays marking its one-hundredth birthday. Nancy Cook's contribution is a cultural glossary that combines government documents with poetry, fiction, and history to examine the role of language in the complex history of water in the American West. Her entry for the term "reclamation" notes that the word remains undefined in the glossary of the USBR, despite its presence in the bureau's name.[1] My goal here is to retain the spirit of Cook's glossary while arguing that proponents of government "reclamation" have often left the term usefully vague in American culture while employing it as justification for immense changes to the North American landscape. The word's ability to hide in plain sight is due in part to its spelling and grammar. "Reclamation" is a nominalization—a noun form of a verb. In this case, the verb is "reclaim." The disappearance of the "i," along with the blunting effect of nominalization, masks the implications of what it might mean to reclaim land and water. While "reclamation" as a term is both vague and bureaucratic, the land management work carried out under reclamation's banner has reimagined geology, hydrology, agriculture, utility, and the purpose of land itself.

In the eighteenth century, colonial English writers used "reclamation" to describe the civilizing of "savages" and "barbarians." In the early nineteenth century, writers in England and the United States began to use the term to describe "the conversion of wasteland . . . into land fit for use."[2] Despite the term's relatively recent evolution, the idea of redeeming undesirable land has been present in American thought for centuries. In *Reinventing Eden*, Carolyn Merchant explains that much of America's thinking about land draws on a Euro-American belief that nature can be "reclaimed" as a biblical, paradisiacal garden.[3] Merchant argues that this "recovery narrative" shaped American thinking about settler-colonial expansion.[4] According to the historian Donald Pisani, the narrative of redemption through reclamation was present in political discussions related to the arid lands in the western United States during the Progressive era:

> Once the best land on the Great Plains had been taken, the United States resorted to a form of continental imperialism—reclaiming the

DOI: 10.4324/9781003581185-5

> arid and semiarid West. Reclamation, it was hoped, would save wasted lives along with wasted land; it would drain the crowded tenements of eastern cities as it salvaged the remaining arable land within the public domain. It would resuscitate the ideal of the family farm, revive civic virtue, and serve as an antidote to a variety of social diseases associated with industrialization and urbanization.[5]

Pisani goes on to link the arguments around reclamation to Frederick Jackson Turner's frontier thesis—another pillar of Progressive-era thinking. For Pisani, the Reclamation Act in 1902—the legislation that created the USBR and was ultimately used to fund the construction of hundreds of dams and irrigation systems—was an attempt to extend the frontier by conquering "wasted land," opening new land for settlement, and preserving the "safety valve" of "free land" that many white Americans felt was necessary to diffuse the class warfare threatening to boil over at the end of the Gilded Age.[6] In this usage of "reclamation," the idea that deserts are environmentally and economically deficient is taken as self-evident, in part because deserts are seen as antithetical to the images of a paradisiacal garden. Thus, "waste spaces" (deserts, arid plains, and sometimes swamps and wetlands) are reclaimed to restore the ideal of paradise, embodied by green agricultural landscapes. This view of deserts as unusable waste spaces lives on in plans to make desert lands in the American southwest "useful" by covering them with solar farms, wind farms, and data centers.[7] For a further examination of American relationships with deserts, see Gary Reger's keyword chapter "Aridity" in this volume.

While political debates surrounding government reclamation were influenced by the garden narrative and the frontier thesis, Donald Worster's *Rivers of Empire* argues that the reclamation of arid lands was primarily about wealth and the domination of nature.[8] Worster notes that congressional arguments about reclamation often reasoned that to leave Western rivers undammed and Western lands unirrigated would result in unclaimed national wealth.[9] Worster's argument is strengthened by the results of the Reclamation Act: nearly a century of excessive dam building, Indigenous dispossession, and the rise of corporate agribusiness in the American West, perhaps most vividly symbolized by the dedication of Hoover Dam in 1935, 30 miles from what was then the small town of Las Vegas, Nevada. In the 1950s, the USBR temporarily adopted the slogan "total use for greater wealth."[10] As of 2025, the bureau has a budget of more than a billion dollars, manages nearly eight million acres of land, and oversees almost 500 dams that provide water for 10 million acres of farmland and 15% of the nation's hydroelectric power. The bureau often refers to itself simply as "Reclamation."[11]

However, interpretations of "reclamation" are not limited to the hegemonic version adopted by the US government. From the late nineteenth century to the present, writers have imagined "reclamation" in concert with their own relationships to North American lands. In *The Conquest of Arid America*

(1905), the irrigation booster William E. Smythe writes that westward expansion "paused" when it encountered aridity and required "reclamation" (a term which appears in his book more than 90 times) to "drive the desert back inch by inch."[12] But Smythe also uses an oxymoron to describe this process, claiming that reclamation is "a policy of peaceful conquest."[13] Smythe's hedging reads as an attempt to ease settler anxieties about the transformation of desert lands. Such anxieties are pronounced in the work of Mary Hallock Foote. Foote's husband Arthur was an irrigation engineer, and much of her fiction in the 1890s is set amid Idaho irrigation projects and dam-building efforts. As I have written elsewhere, Foote's Idaho fiction adopts gothic conventions to express uneasiness about reclaiming a landscape she found forbidding.[14] For Foote, reclamation is a "violent and promiscuous birth"[15] that relies on captive women and captive water to serve as mothers, birthing a new landscape in a process that destroys as it creates.

Writing almost a century later, Wendell Berry sees "reclamation" as a process of restoration rather than creation. In "The Making of a Marginal Farm" (1980), Berry describes a project to rehabilitate 40 acres of land abandoned by a developer. "What we had bought," Berry writes, "was less a farm than a reclamation project."[16] Berry's usage adheres to the nineteenth-century definition of "reclamation," but his underlying assumptions do not resemble the extractive values of "total use for greater wealth." The essay argues that "marginal" land is the result of failed human values and a history of misuse rather than innately deficient characteristics of the land itself: "they were not 'marginal' because they ever were unfit for human use, but because in both culture and character *we* had been unfit to use them."[17] A similar idea is present in the recent usage of "reclamation" to describe efforts to rehabilitate and restore exhausted mining sites.[18]

Human relationships with land are also central to reclamation as the concept relates to the Indigenous peoples of North America. One goal of the federal government's reclamation efforts has been to wrest control of land and water from Native Americans while attempting to erase Indigenous histories, which do not define these lands as waste spaces in need of redemption. For much of the twentieth century, the USBR often worked with the Bureau of Indian Affairs. Pisani notes that the two agencies saw their goals—the colonization of the land and the colonization of the Native Americans—as linked: "leaders in the Reclamation Service recognized that transforming 'savage' Indians into peaceful farmers had no less appeal than transforming the wilderness into a garden."[19] This ideology is exemplified by the Pick-Sloan Missouri Basin plan—coauthored by the Army Corps of Engineers and the USBR (the two most prolific dam-building organizations in the US government)—which built dams on the Missouri River in the mid-twentieth century and flooded portions of several reservations, including 200,000 acres of the Standing Rock and Cheyenne River Reservations in South Dakota.[20] Unsurprisingly, signifiers of government reclamation and dam building often appear in Indigenous literature. See, for example, *Wind from an Enemy Sky*

(1978) by D'Arcy McNickle (Salish Kootenai), which depicts the construction of a dam on a fictional reservation. Additionally, Jessica Cory's chapter in this volume explores *Solar Storms* (1995) by the Chickasaw writer Linda Hogan, a novel focused on dam construction along the border between the United States and Canada.

However, Indigenous and environmental activists are reclaiming "reclamation" as a decolonial tool. Movements toward the restoration of languages, ecosystems, sovereignty, and homelands in North America use "reclamation" for a variety of political, environmental, and cultural purposes.[21] In addition, Indigenous movements have adopted another signifier that leverages useful ambiguity: "Land Back." Appearing on social media in the late 2010s and credited to Arnell Tailfeathers (Blackfoot), Land Back (or #landback) is a rallying cry for decolonization and the rise of movements to reclaim stewardship over tribal homelands.[22] The phrase's ambiguity is highlighted by a scene in the television series *Reservation Dogs*. An episode titled "Uncle Brownie" opens with an elderly white couple driving past a sign spray-painted with the word "LANDBACK." The man is flummoxed by the implications. "What do you suppose that means?" he asks.[23] Here, the ambiguities of reclamation are reversed. Reclamation has always hidden behind a shroud of obfuscation. We recognize the word, but it seems almost divorced from its potential meanings. "Land Back" asks us to reunite word and meaning, asks us to return the land to ecological health and Indigenous stewardship. "Land Back" suggests that we take reclamation seriously as an idea, that we consider who has a rightful claim to "reclaim" land while honoring the claims of the land itself.

In a recent essay, environmental humanists Jacob Goessling and Jordan B. Kinder argue that reclamation can be seen as a way of reconstructing the relationship between humans and land by referring to "a range of approaches from social-justice oriented modes of care and repair to the kinds of Indigenous reclamation at the center of growing demands for Land Back."[24] In other words, Land Back, environmental justice, and environmental restoration represent important attempts to reclaim reclamation from extractive forces that see land primarily as a means of profit-making. To reclaim reclamation is to envision new ways of thinking about an old idea that has historically been misused by colonial forces yet holds the potential for reimagining the relationships between North Americans and the lands they inhabit.

Notes

1 Cook, "Writing Water in the West: Reclaiming the Language of Reclamation," 851–862.
2 *Oxford English Dictionary*, s.v. "reclamation (n.)," July 2023, https://doi.org/10.1093/OED/6432344522.
3 Merchant, *Reinventing Eden*, 3.
4 Merchant, *Reinventing Eden*, 7.

5 Pisani, *Water and the American Government*, xii.
6 Deverell, "To Loosen the Safety Valve," 269–285.
7 Wainright, "How Solar Farms Took Over."; Peterson, "Solar Project to Destroy."
8 Worster, *Rivers of Empire*.
9 Worster, *Rivers of Empire*, 165–169.
10 Worster, *Rivers of Empire*, 266.
11 Bureau of Reclamation, "About Us."
12 Smythe, *The Conquest of Arid America*, xv.
13 Smythe, *The Conquest of Arid America*, 8.
14 Grover, "Mary Hallock Foote's Idaho Fiction," 21–50.
15 Foote, *The Last Assembly Ball*, 7.
16 Wendell Berry, *The World-Ending Fire*, 41.
17 Wendell Berry, *The World-Ending Fire*, 41. Emphasis in original.
18 Goessling and Kinder, "Reclamation," 236–239.
19 Pisani, *Water and the American Government*, 154–155.
20 Lawson, *Dammed Indians*.
21 Atleo and Boron, "Land Is Life"; Ferguson and Weaselboy, "Indigenous Sustainable Relations"; Leung, "An 'Othered' Land Reclamation"; Moscufo, and Trujillo, "These Indigenous Artists." These sources represent a varied set of examples of this usage of "reclamation."
22 Yesno, "Land Back."
23 Harjo, et al., "Uncle Brownie."
24 Goessling and Kinder, "Reclamation," 237.

Bibliography

Atleo, Clifford, and Jonathan Boron. "Land Is Life: Indigenous Relationships to Territory and Navigating Settler Colonial Property Regimes in Canada." *Land* 11, no. 609 (2022). https://doi.org/10.3390/land11050609.

Berry, Wendell. *The World-Ending Fire: The Essential Wendell Berry*. Counterpoint, 2017.

Bureau of Reclamation. "About Us." *Facts | Bureau of Reclamation*. Accessed September 18, 2024. https://www.usbr.gov/main/about/fact.html.

Cook, Nancy. "Writing Water in the West: Reclaiming the Language of Reclamation." In *The Bureau of Reclamation: History Essays from the Centennial Symposium Volumes I and II*, 851–862. Bureau of Reclamation, U.S. Department of the Interior, 2008.

Deverell, William F. "To Loosen the Safety Valve: Eastern Workers and Western Lands." *Western Historical Quarterly* 19, no. 3 (August 1988): 269–285.

Ferguson, Jenanne, and Marissa Weaselboy. "Indigenous Sustainable Relations: Considering Land in Language and Language in Land." *Current Opinion in Environmental Sustainability* 43 (2020): 1–7.

Foote, Mary Hallock. *The Last Assembly Ball, and The Fate of a Voice*. Houghton, Mifflin and Company, 1889. Hathitrust Digital Library. http://hdl.handle.net/2027/uc2.ark:/13960/t0ns0mh94.

Goessling, Jacob, and Jordan B. Kinder. "Reclamation." *Environmental Humanities* 15, no. 2 (2023): 236–239.

Grover, Quinn. "'Taken from Their Self-Found Paths': Captivity and Creation in Mary Hallock Foote's Idaho Fiction." *Western American Literature* 57, no. 1 (Spring 2022): 21–50.

Harjo, Sterlin, Taika Waititi, and Sydney Freeland, writers. *Reservation Dogs*, Season 1, episode 3, "Uncle Brownie." Directed by Blackhorse Low, featuring Devery Jacobs, D'Pharaoh Woon-A-Tai, and Lane Factor. 2021, *Hulu*. hulu.com.

Leung, Arthur. "An 'Othered' Land Reclamation: Decolonization in Anticipation of Another Great Flood." *Journal of Architectural Education* 75, no. 2 (September 2021): 170–183. https://doi.org/10.1080/10464883.2021.1947671.
McNickle, D'Arcy. *Wind from an Enemy Sky*. University of New Mexico Press, 1988.
Moscufo, Michela. "For These Indigenous Artists 'Land Back' Is Both a Political Message and a Fundraising Opportunity." *Forbes*, September 29, 2020. Accessed August 27, 2024. https://www.forbes.com/sites/michelamoscufo/2020/09/29/for-these-indigenous-artists-land-back-is-both-a-political-message-and-a-fundraising-opportunity/.
Oxford English Dictionary, s.v. "Reclamation (*n.*)." July 2023. https://doi.org/10.1093/OED/6432344522.
Peterson, Melody. "Solar Project to Destroy Thousands of Joshua Trees in the Mojave Desert." *Los Angeles Times*, May 31, 2024. Accessed September 22, 2024. https://www.theguardian.com/us-news/2023/may/21/solar-farms-energy-power-california-mojave-desert.
Pisani, Donald J. *Water and American Government: The Reclamation Bureau, National Water Policy, and the West, 1902–1935*. University of California Press, 2002.
Smythe, William E. *The Conquest of Arid America*. New and Revised ed. The Macmillan Company, 1905.
Trujillo, Simón Ventura. "'So That the Thieves Will Not Inherit the Earth': Writing and the Fugitive Translation of Indigenous Land Reclamation." *Critical Ethnic Studies* 3, no. 1 (Spring 2017): 59–89.
Wainright, Oliver. "How Solar Farms Took Over the California Desert." *The Guardian*, May 21, 2023. Accessed September 22, 2024. https://www.theguardian.com/us-news/2023/may/21/solar-farms-energy-power-california-mojave-desert.
Worster, Donald. *Rivers of Empire: Water, Aridity, and the Growth of the American West*. New Pantheon, 1985.
Yesno, Riley. "Land Back." *New Internationalist*, October 24, 2022. Accessed August 27, 2024. https://newint.org/features/2022/10/24/land-back-decolonize-how.

4 Allotment

Jonathan Radocay

The transformation of Indigenous land relations under settler-colonial pressure has profoundly shaped the evolution of land, water, and resource management in North America. Over the course of this colonial history, settler legislation and law have impacted nearly every aspect of Native life, including sovereignty and nationhood; access to and stewardship of customary hunting, fishing, and gathering places; family and community integrity; and the transmission of language and culture from one generation to the next. In the United States, few pieces of legislation have had such a singular impact as the Dawes Act of 1887, which inaugurated an era of US federal Indian policy known as allotment. The Act established provisions to forcibly privatize reservations and other Indigenous lands held in common and to confer US citizenship to allottees who received individually owned parcels of land called allotments. Through private landownership, the Dawes Act offered a sweeping colonial vision of Native racial uplift and assimilation based on settler principles of privatized land management and economic development.

Its actual administration, however, rarely followed this dubious vision: lands often ended up in settler hands or under state ownership through the fraudulent coercions of expanding state jurisdiction over tribal lands and a settler-dominated land market.[1] Subsequent legislation amending the Dawes Act, including the Burke Act of 1906, introduced separate competency requirements and other racial restrictions for so-called Indian landowners.[2] These amendments tied US citizenship and Native political rights to racist assumptions about the ability of Native people to manage land. Even as the allotment movement fervently argued for privatization as a means to liberate Native people from the federal paternalism of the agency reservation system that defined much of nineteenth-century federal Indian policy,[3] these amendments introduced new racialized bureaucratic restrictions that reaffirmed paternalism and placed limits on leasing, land use, and the ability of allottees to access credit and other financial resources otherwise available to non-native landowners. Despite being pivotal to redrawing the boundaries of tribal, public, and private land management in the United States, scholarship on US environmental history and federal Indian policy has often positioned the Dawes Act and allotment policy at the margins of land management history.

DOI: 10.4324/9781003581185-6

Even when allotment is covered, its legacy is often presented through narratives of dispossession that minimize Indigenous agency.

From a Native perspective, allotment and the colonial origins of private land markets across North America have been fundamental to the transformation of land relations—not simply a footnote in the long history of settler land possession and development. In this keyword entry, I argue that allotment is not just central to understanding the imbrication of settler colonialism, federal Indian policy, and the development of US land management policies and practices. It has also set in motion fundamental, and often overlooked, transformations in Indigenous land relations and sovereignty. While this entry aims to establish allotment as a key concept in land management discourse about land tenure, ranching and agriculture, water rights, and oil extraction across the United States, I show how these concerns are inextricable from past, present, and future concerns related to Indigenous nation-building, environmental stewardship, and land management practices. Presenting a range of Indigenous writing, fiction, and creative works produced before, during, and after the Dawes Act, the entry grounds literatures of land management within Indigenous histories, land relations, and experiences with allotment. Across many different forms, texts, and genres, Indigenous literatures have served as important sites for developing incisive critiques of colonial privatization and its racialized, gendered violence. More importantly, these literatures articulate Indigenous futures amidst this violence and offer blueprints for envisioning management practices, policies, and relations that continue to serve Native Nations and communities in a post-allotment world.

Settler literature and media have played an important role in moving allotment to the status of historical footnote. These texts often redraw Indigenous engagement with allotment within literary representations of frontier spaces, in which Native Nations and Indigenous communities have already been eliminated or marginalized. Willa Cather'srepresentations of settler agricultural life in her regionalist novels *O Pioneers!* (1913) and *My Antonia* (1918) reproduce colonial land relations in the Nebraska region, while erasing Omaha experiences with allotment and forced Removal.[4] In Western popular fiction and media, including novels by Owen Wister and Zane Grey, stories published in pulp magazines like *Western Story* and *Cowboy Stories,* and blockbuster films directed by Raoul Walsh, John Ford, and Howard Hawks, the priorities of running a ranch—its cattle driving, division of labor among ranch hands, and settler domesticity—marginalize the history of allotment policy that created the privatized land relations underlying the narratives themselves. This media imagines the land only within a singular, often invisible, frame of land relation, as always privatized, just waiting for economic development and management by enterprising pioneers.

Named after one of its principal architects, Massachusetts Senator Henry L. Dawes, the Dawes Act laid out four main provisions that would guide colonial privatization in the United States for nearly five decades: (1) land grants held in severalty by individual tribal members; (2) racial restrictions on these allotments for a number of years (usually 25) based on blood quantum; (3) an ill-defined process by which allotment would be administered; and (4) US citizenship.[5] These provisions show the parallel (il)logics of dispossession and privatized settler land development with which Native Nations contended. Under this policy, territorial expansion onto Indigenous lands would function through widening the relations of agricultural yeomanry and private resource extraction: through private management, Indigenous lands would be cultivated and incorporated into US territory, subject to laws governing title and land use. Both land and person would be managed within privatized land relations, even as paternalistic bureaucracy, law, and racial restrictions barred Indigenous people from fully participating within these newly redrawn boundaries of land use and development.[6] For all the promises that allotment may have held for individual tribal members, particularly those from mixed, white elite families, forced privatization made other members, especially from Black and so-called full-blood communities, more vulnerable to fraud and the manifold forms of violence introduced by logging, ranching, mining, and other privatized economies of extraction.

From the early days of its policy advocacy into our present, post-allotment landscape, Indigenous authors both contested the paternalistic, fraudulent nature of its administrative processes and imagined new possibilities for nation-building and Indigenous land relations. During the years leading up to the Dawes Act, Sarah Winnemucca (Numa [Northern Paiute]) advocated for a Native-administered allotment policy as a means of protecting Native women and girls from the sexual violence of ranching and cattle interests.[7] Albeit with significant qualification, Winnemucca saw that private land tenure offered more spatial autonomy for Numa communities than the agency reservation system, which operated through a colonial tenure system that held public federal land "in trust" for the supposed benefit of Native Nations.[8] In a petition included at the end of her autobiographical narrative, *Life Among the Piutes: Their Wrongs and Claims* (1883), she calls for Numa people imprisoned on the Yakima Reservation to be restored to the Malheur Reservation located in what is now known as Eastern Oregon.[9]

Although *Life* spans from the 1840s to the 1880s, decades preceding the allotment policy discourse with which she engaged, Winnemucca writes nearly a half century of Numa history with settler encounter as a citation to support the allotment petition that concludes the narrative. The petition orients the narrative within contemporary allotment policy discourse and toward an important political goal: Winnemucca saw allotment as a strategic means by which land could be restored and relatives could be returned—but not within the agency reservation system she critiques in the preceding narrative. Throughout *Life*, Winnemucca demonstrates that the reservation system has resulted in the illegal expansion of settler cattle and ranching interests onto reservation lands, which has both threatened Native women and girls and Numa food systems of fishing, camas root harvesting, and pine nut collection, in which women and girls play a central role. She imagines the inadvertent possibilities that allotment's provisions, which would be codified four years later, might have for protecting Native women and girls from the sexual violence of settler land development[10] and for extending their vital role in Indigenous land relations, including but not limited to Numa food systems.

While Winnemucca saw potential for tribal relations in allotment, poet and newspaper editor Alexander Posey (Muscogee [Creek] Nation) considered the dangers that privatization posed to Indigenous land tenure systems in Indian Territory. On March 1, 1894, in the *Indian Journal*, he published a parody of Shakespeare's famous soliloquy in *Hamlet*. In this poem, he satirizes the allotment policy discussion in the press:

> To allot, or not to allot, that is the
> Question; whether 'tis nobler in the mind to
> Suffer the country to lie in common as it is,
> Or to divide it up and give each man
> His share pro rata, and by dividing
> End this sea of troubles? To allot, divide,
> Perchance to end in statehood;
> Ah, there's the rub![11]

Posey maps the same existential contemplation that Hamlet performs onto the stakes of allotment to Creek sovereignty. The verb for the privatization of Creek lands, "To allot," substitutes for the original existential linking verb, "To be." This parodic substitution conveys that the process of privatization, "to allot," amounts to an existential transformation of Creek sovereignty and land relations. Moreover, by parodying Hamlet's contemplation of suicide, Posey ironically frames the allotment policy being debated in the press as a question of Creek national survival.

Although the Dawes Act was passed seven years prior, Indian Territory was initially exempted due to successful lobbying by the Five Tribes, including Posey's Muscogee (Creek) Nation.[12] However, the pressure to privatize *all* of Indian Country remained, especially by oil, coal, and timber interests that wanted to acquire mineral and land rights. As Posey's poem indicates, the policy's extension over Indian Territory remained an active topic of land management concern in the press. Many settler pro-allotment positions turned on discourses of racial uplift through economic development.[13] Through these debates, satirized in Posey's poem, Indian Territory print culture makes visible the profound differences in US and Native land tenure systems. These debates center questions of land tenure, treaty rights, and sovereignty to US national concerns of economic development and land management in ways that were not necessarily evident outside of Indian Territory. Four years after Posey's parody, Congress would successfully pass the Curtis Act of 1898 that extended allotment—albeit unevenly—to Indian Territory.

Although millions of acres of Indigenous land would be allotted following the Dawes Act, Curtis Act, and subsequent legislation, Indigenous authors continued to critique allotment's immediate impact and to imagine new futures within its profound transformations. Cree-Métis writer D'Arcy McNickle's (Confederated Salish and Kootenai Tribes) posthumous novel, *Wind from an Enemy Sky* (1988),[14] critiques hydrological development made possible only through the privatization of lands belonging to the Little Elk, a fictionalized Native Nation resembling one belonging to the Confederated Salish and Kootenai Tribes of the Flathead Reservation. The novel turns on conflict between Little Elk and US settler interests over the construction of a dam much like the Kerr Dam (now known as Seli'š Ksanka Qlispe' Dam) constructed on the Flathead Reservation. The novel restores Indigenous experiences with allotment—and its abrogation of previous treaty rights—as being fundamental to the development of US infrastructure and water management, undercutting the well-intentioned visions of liberal industrialists like Adam Pell who finances the dam in the novel. Much like Posey does with land tenure systems, McNickle unsettles settler notions of water as a resource to be enclosed for national economic development by making visible Indigenous relationships to streams, rivers, and other waterways that prioritize the human, nonhuman, and other-than-human worlds water sustains.

Even after allotment policy officially ended with the Indian Reorganization Act of 1934, which halted the Dawes Act and established pathways to restore land to tribal control, the policy has continued to impact Native Nations and Indigenous communities. The contributors to *Allotment Stories* (2021)[15] look to narrative forms, storytelling, and other methods traditionally marginalized from scholarship on allotment and land management. From her own family's experience with returning their individual allotments to tribal control, Sheryl Lightfoot (Anishinaabe, Lake Superior Band) considers the challenges presented by the Cobell Land Buy-Back Program, which enables Native Nations to purchase fractionated interests from allottee heirs for the purpose of tribal land development. However, she shows that for families like hers, this program can be an agonizing and emotional process.[16] This buy back program is set up to address title fractionation. As an allotment passes down through the generations, land title is divided into increasingly numerous and tinier fractions, impeding descendants from benefitting from leasing and other opportunities of economic development. Sometimes, this title fractionation even pits individual members against tribal governments that move to transfer former allotment lands back into trust status.[17]

In a creative-critical piece, Sarah Biscarra Dilley (yak titʸu titʸu yak tiłhini Northern Chumash Tribe) uses poetry, art, photography, and documentary material to confront the modern-day legacies of land privatization, ranching, petroleum extraction, and other forms of land loss in California through the experiences of her maternal relatives. Dilley layers maps depicting contemporary land-use activity in California, including gas and oil fields and carbon sequestration, with a map of 18 unratified treaty areas. She shows that these economies of resource extraction and sequestration often take place on treaty lands, including her own family allotment parcels and ancestral villages.[18]

Although the so-called allotment era references a period from 1887 to 1934, the impact of colonial privatization and land management practices on Native Nations far exceeds this narrow timeframe or even the political boundaries of the United States itself. Then as now, privatization schemes like allotment have shaped the contours of public, private, and tribal management of land and water. Importantly, Native Nations and Indigenous communities across North America continue to reclaim lands, revitalize language and culture, and restore communities impacted by privatization and enclosure by drawing on Indigenous knowledges, community histories, and other innovative approaches to environmental stewardship. For example, in Mexico, Indigenous peoples create primordial land titles through community history "mosaics" that combine storytelling from Elders and visual media. Through stories of enduring presence, kinship relations, and other lived experiences, these subversive texts support ongoing land claims in court and remap privatized landscapes in dispute within vital Indigenous worlds.[19] In the United States and Canada, Native Nations confront challenges presented by the patchwork of tribal, public, and private land tenure systems left behind by colonial privatization through the creation of tribal national parks that differ from park systems based on settler conservation and management priorities. These tribal parks reterritorialize privatized land and reassert Indigenous sovereign space, balancing priorities related to culture and ceremony, including restricting access to sacred sites, with those related to tourism, small-scale harvesting activities, and other forms of tribal economic development.[20] Outside of legal venues and government-to-government relationships, Native Nations also engage land trusts, conservancies, and other private conservation tools to reclaim privatized land and to protect the sustainability of cultural resources. Tribes use these tools to create partnerships with non-native landowners, non-profits, and other stakeholders within the world of private land conservation.[21]

Indigenous struggles over land and water, and against racialized and gendered violence, demonstrate that these issues are not simply reducible to conservation and management concerns. They are inseparable from Indigenous histories, experiences, and knowledges. This short entry establishes a brief and certainly not exhaustive genealogy of Indigenous allotment politics, history, and literature, from the early days of policy discourse before the Dawes Act to our contemporary post-allotment present. This genealogy shows that the diverse ways Indigenous authors and Native Nations engage with and confront allotment are integral to broader histories of land and resource management throughout North America. First and foremost, however, these engagements and confrontations represent a long tradition of Indigenous nation-building efforts, environmental stewardship, and healing from the forms of severance allotment has introduced. In this tradition, we find priorities rooted in maintaining family and community integrity, cultivating good relations with land and water, and sustaining cultural resources into the future.

These values continue to guide Native Nations and Indigenous communities through the post-allotment world we have all inherited.

Notes

1 For a detailed case study of allotment's administration in Indian Territory, see Debo, *And Still the Waters Run.*
2 U.S. Congress, *General Allotment Act.*
3 For an account of the allotment movement's rhetoric and political justifications, see Otis, *The Dawes Act*, 8–32.
4 Strimple Fuisz, "Mapping and (Re)mapping."
5 U.S. Congress, *An Act to Provide.*
6 Ruppel, *Unearthing Indian Land*, 6.
7 Zanjani, *Sarah Winnemucca*, 243.
8 Other Native Nations also engaged with the allotment process as a means of securing greater spatial autonomy. See Greenwald, *Reconfiguring the Reservation* and Tonkovich, *The Allotment Plot.*
9 Winnemucca, *Life Among the Piutes*, 247.
10 For a discussion of how Winnemucca's writing underlines the inextricable connections between sexual violence and land dispossession, see Deer, *The Beginning and End of Rape*, 117–118.
11 Posey, *Song of the Oktahutche*, 26.
12 For more detail on this exemption, see Carter, *The Dawes Commission*, 1–2.
13 For a compendium of information about over 200 American Indian and Alaska Native newspapers and periodicals, see Littlefield and Parins, eds., *American Indian and Alaska Native Newspapers and Periodicals.* For details about how the settler press influenced the passage of the Dawes and Curtis Acts, see Coward, *The Newspaper Indian*, 196–226.
14 McNickle, *Wind from an Enemy Sky.*
15 Heath Justice and O'Brien, eds., *Allotment Stories.*
16 Lightfoot, " 'What Should We Do?'."
17 For a larger discussion of fractionation and its enduring problems, see Ruppel, *Unearthing Indian Land*, 32–35.
18 Dilley, "tʸiptukiɬhɨ wa tʸiptutʸɨʔnɨ."
19 McDonough, "Settler Colonial Mexico."
20 Carroll, "Native Enclosures," 31–40.
21 Middleton, *Trust in the Land.*

Bibliography

Carroll, Clint. "Native Enclosures: Tribal National Parks and the Progressive Politics of Environmental Stewardship in Indian Country." *Geoforum* 53 (2014): 31–40.
Carter, Kent. *The Dawes Commission and the Allotment of the Five Civilized Tribes, 1893–1914.* Ancestry.com, 1999.
Coward, John. *The Newspaper Indian: Native American Identity in the Press, 1820–1890.* University of Illinois Press, 1999.
Debo, Angie. *And Still the Waters Run: The Betrayal of the Five Civilized Tribes.* Princeton University Press, 1972.
Deer, Sarah. *The Beginning and End of Rape: Confronting Sexual Violence in Native America.* University of Minneapolis Press, 2015.
Dilley, Sarah Biscarra. "tʸiptukiɬhɨ wa tʸiptutʸɨʔnɨ, Where Are You from and Where Are You Going?" In *Allotment Stories*, edited by Daniel Heath Justice and Jean M. O'Brien. University of Minnesota Press, 2021.

Fuisz, Lisbeth Strimple. "Mapping and (Re)mapping the Nebraska Landscape in the Works of Willa Cather and Francis La Flesche." In *Cather Studies: Volume 14*, edited by Ann Romines and Marilee Lindermann. University of Nebraska Press, 2025.

Greenwald, Emily. *Reconfiguring the Reservation*. University of New Mexico Press, 2002.

Justice, Daniel Health, and Jean M. O'Brien, eds. *Allotment Stories*. University of Minnesota Press, 2021.

Lightfoot, Sheryl. "'What Should We Do?' Returning Fractionated Allotments Back to the Tribes—One Family's Story." In *Allotment Stories*, edited by Daniel Heath Justice and Jean M. O'Brien. University of Minnesota Press, 2021.

Littlefield, Daniel, and James Parins, eds. *American Indian and Alaska Native Newspapers and Periodicals*. Greenwood Press, 1984.

McDonough, Kelly S. "Settler Colonial Mexico and Indigenous Primordial Titles." In *Allotment Stories*, edited by Daniel Heath Justice and Jean M. O'Brien. University of Minnesota Press, 2021.

McNickle, D'Arcy. *Wind from an Enemy Sky*. University of New Mexico Press, 1988.

Middleton, Beth Rose. *Trust in the Land: New Directions in Tribal Conservation*. University of Arizona Press, 2011.

Otis, D.S. *The Dawes Act and the Allotment of Indian Lands*. University of Oklahoma Press, 1973.

Posey, Alexander. *Song of the Oktahutche*, edited by Matthew Wynn Sivils. University of Nebraska Press, 2008.

Ruppel, Kristin T. *Unearthing Indian Land: Living with the Legacies of Allotment*. University of Arizona Press, 2008.

Tonkovich, Nicole. *The Allotment Plot: Alice C. Fletcher, E. Jane Gay, and Nez Perce Survivance*. University of Nebraska Press, 2012.

U.S. Congress. *An Act to Provide for the Allotment of Lands in Severalty to Indians on the Various Reservations (General Allotment Act or Dawes Act)*. Statutes at Large 24, 388–91, 1887.

U.S. Congress. *General Allotment Act Amendment of 1906 (Burke Act)*. Statutes at Large 34, 59–149, 1906.

Winnemucca, Sarah Hopkins. *Life Among the Piutes: Their Wrongs and Claims*. University of Nevada Press, 1994.

Zanjani, Sally. *Sarah Winnemucca*. University of Nebraska Press, 2001.

5 Islandness

Sreya Ann Oommen

Introduction

Histories of islands are riddled with preconceptions about islands as finite, malleable spaces that made them sites of various, often aggressive and extractive, colonial pursuits. Godfrey Baldacchino, a pioneer of island studies who was instrumental in developing island-based theories of islandness, notes that the misrepresentation of islands as "microcosms of mainlands" is a major trap that island researchers should avoid.[1] However, the liminal position of these landmasses as spaces excluded from universal visions of development, even when they are spaces that are most affected by climate disasters, makes them all the more important to consider as we continue imagining more just and sustainable land management futures.[2] Reclaiming narratives about island spaces is crucial in the Anthropocene where islands are sites most affected by climate change, as elucidated by the increased magnitude of lives lost due to hurricanes, typhoons, and other natural disasters. These natural disasters are an indirect consequence of anthropogenic activities and reveal particular vulnerabilities among the populations who live in close proximity to oceans. Additionally, colonial land management patterns often viewed island spaces as "pristine, exotic, manageably simple social microcosms or physical laboratories."[3] A direct consequence of centuries of colonial exploitation includes the non-sustainable colonial practice of monocropping agriculture in plantations, which diminishes the fertility and health of the soil.[4] Caribbean island communities such as Jamaica, which is the primary focus of this keyword entry, settled into monoculture by the late seventeenth century, and this led to "a sugar and slave system in its starkest and most exploitive[sic] form."[5] Within the contexts of the Anthropocene and decoloniality, I intervene in the discourse of islandness to explore how microhistories of islandness become useful tools for reimagining land management and sustainable futures.[6]

Islandness, the Anthropocene, and Decoloniality

What is islandness? How can the use of this term when we envision collective global futures offer us sustainable ways of managing land in the

DOI: 10.4324/9781003581185-7

Anthropocene? Island studies scholar David Weale elucidates islandness as an integral aspect of one's being, "a part as deep as marrow, and as natural and unselfconscious as breathing. . . . [I]slandness, is the source and reference point for the imagination of Islanders."[7] The deeply rooted sense of being on an island affects islanders in all aspects of their lives and acts as a fundamental source of their "communal insight and wisdom."[8] Centuries of colonial history positioned islands as places of mystery, paradox, paradise, and laboratories, among other identities. Islandness considers islands within history and literature to be spaces that assimilate "myth and lore," "sacred and profane," "respite and danger," and "reality and fantasy, utopia and dystopia, isolation and connection," among several other binaries.[9] In other words, islands are not finite but are, instead, capacious enough to contain everything. In his book *Islandology*, Marc Shell notes that islandness, in terms of identity, "informs primordial issues of philosophy: how, conceptually, we connect and disconnect parts and wholes."[10] Baldacchino prefers "islandness" to "insularity," a word associated with the "semantic baggage of separation and backwardness."[11] The word "islandness," a "more neutral choice," does not subject islands, islanders, and other island species to "negative assessment" in the Anthropocene and expands the accommodability of the field of island studies.[12]

An increasing number of island studies scholars prefer the term "islandness" to "represent all that constitutes an island; a distilled and judgement-free sense of island living."[13] As islandness encompasses plural perspectives, it expands the potential of decolonial island studies that dissociate island spaces from ascribed meanings and perceptions. Islandness encompasses how islanders embrace their roots, traditional practices, land, and means of life; furthermore, islandness accounts for those communities which integrated into island spaces at different points in history and might struggle to thrive while navigating the consequences of their complicated pasts. Islandness also moves away from the ascribed notions of islands as finite and controllable spaces. Brodber's fictional depiction of Jamaican islandness that is deeply rooted in the traditional African Jamaican practices, where humans and nature exist as part of each other, highlights the significance of islands within decoloniality in the Anthropocene. By considering the impact of human activities on the environment, Paul J. Crutzen, who popularized the idea of the Anthropocene, finds the term appropriate to emphasize humankind's pivotal and damaging, but not entirely irredeemable, role in impacting geology and ecology.[14]

Island studies scholars emphasize the importance of amplifying island voices that dominant narratives often overlook, recognize the need for sustainability in a world altered by human activity, and highlight the relevance of sustainable practices in island communities. The ongoing movement in Hawaii to resurrect the islands' traditional *ahupua'a* and *konohiki* systems to "better care for Hawaii's natural resources and residents" is an instance where islanders depend on traditional, sustainable methods of managing and preserving land that is part of their inherent islandness.[15] Elizabeth DeLoughrey highlights the significance of listening to island narratives when considering sustainability that transcends hierarchical structures in the Anthropocene:

> [in] an era of a truly global environmental crisis, Anthropocene scholarship cannot afford to overlook narratives from the global south, particularly from those island regions that have been and continue to be at the forefront of ecologically devastating climate change.[16]

Many island studies scholars consider decolonial island studies as the discipline's future as it offers a direction of research in fields and regions that, thus far, have received limited attention. Walter D. Mignolo, a prominent decolonial theorist, perceives decoloniality as a form of "epistemic disobedience" and "epistemic delinking" that builds from postcolonial thought.[17] Mignolo elucidates that while decoloniality moves "away and beyond the post-colonial," postcolonial criticism addressed the process of "scholarly transformation within the academy."[18] Decolonial thinking entails continuous reinscription of ways of living that were "violently devalued or demonized by colonial . . . and interventionist agendas."[19] Such agendas include those related to land and oceanic management. The misrepresentation of island spaces as finite and controllable, based on their shapes and sizes, explicates how Western narratives have often silenced island voices. Decolonial studies scholars focus on shifting the narrative away from the colonizer; additionally, they reject a colonial discourse that perceives those formerly colonized only within a framework within which they are linked to the colonizer. The renewed way of thinking focuses on the experiences, voices, and realities of the individual without the colonial matrix of power as a reference, and Mignolo posits decolonial thinking as "one type of imperative practice that aims to delink from the coloniality of knowledge and being (that is, from imperial/colonial subjections of subjectivities through knowledge)."[20]

Brodber's novel *Nothing's Mat* amplifies the decolonial nature of islandness as it represents the delinking from colonial narrative patterns/traditions. Brodber makes her narrative quintessentially African Jamaican by incorporating Jamaican Creole, a remnant of Jamaica's complex colonial history. *Nothing's Mat*, a multivocal narrative, describes the story of an unnamed female protagonist who belongs to a Jamaican family and is born and brought up in England. She visits Jamaica to do research on her West Indian family in her sixth form. The protagonist meets a relative in her paternal grandmother's generation, June Reid, whose pet name is "Nothing." The novel reveals various possible reasons behind the name Nothing that are entangled with the association of islandness with nothingness, chaos, and order. When the protagonist starts her project on the West Indian family, she assumes that all she has to do is to get her parents' respective family trees. The protagonist uses her own family as the base for her research and then examines its structure. She looks at her father's side of the family, and when she tries to fit her family on a family tree using lines and arrows between the individuals, the process often makes her father say, "Then where does X fit in?"[21] Her mother suggests that the protagonist should acknowledge the exercise as futile and admit that theirs is the "alternative family" as they are "going to fit neatly into no mould."[22] The protagonist's mother's words pave the way to the part of the narrative that describes the complexity of the Jamaican family, and her father suggests that the best option for her project would be for her to visit her father's side of the family in Jamaica. When the protagonist interacts with the family in Jamaica, she realizes the impossibility of representing her Jamaican family using the Western system of family trees, the most commonly used structure for understanding and representing families. This realization leads her to make a circular family mat out of locally available materials. The protagonist receives information about her Jamaican family from Nothing, who is revealed to be a first cousin of the protagonist's paternal grandmother. Brodber delinks the conception of a Jamaican family from the Western framework of a family tree and instead uses a mat woven out of *ping wing macca* to represent familial lineage.

In *Nothing's Mat,* Brodber shows how the families reflect islandness specific to the Caribbean by proliferating patterns across generations. Antonio Benítez-Rojo, when he studies the Caribbean, mentions that the history of suppression, indentured labor, and forced migration in the Caribbean archipelago leads to a scenario where chaos coexists with nothingness as he considers the Caribbean to be a meta-archipelago without limits or a center, "a chaos within which there is an island that proliferates endlessly, each a copy of a different one."[23] *Nothing's Mat* offers an illustration of the endless proliferation of patterns in the archipelago and the manner in which islandness delinks from the colonial frameworks imposed on them through a family mat where the patterns represent family traits that proliferate across different generations.

Replicating the pattern in which leaves appear on the plant, the protagonist represents the different members of her family who are related to each other not just by blood but by adoption as well. Here, Brodber demonstrates how island communities, which have often been perceived through colonial frameworks, break from the colonial matrix of power by rooting her family's narrative back to the island's soil and patterns that are found in nature. Hannah Muhlfelder, in her entry in this volume: "Soil Futures: Environmental Management as Care in Mary Ann Shadd Cary's "A Plea for Emigration"," mentions how soil is often considered to be part of "an interconnected system of physical environment and biota."[24] Nothing's family mat illustrates a new way of managing land—the mat ties people to the land and shows the community's interdependence. They are not separate from or hierarchically above the island—they *are* the island. The mat demonstrates a new approach to managing the land the family inhabits. The protagonist's circular family mat, with one central circle and smaller circles arranged in descending order around the main one, allows her to reimagine her family's rootedness to the island. Each of the circles signifies a family member, and the narrative elucidates how the interconnection of the mat strings exemplifies the entangled familial relationships in the African Jamaican community.

Brodber's sociological and literary research always addresses the deep association among Jamaican rural communities, their land, plantations, and histories. The author's quest to develop a space that preserves and continues African Jamaican traditions is reflected through the protagonist's interest in her family's traditional practices, including agricultural practices, as she looks for research data in unconventional places, such as oral recollections of history by various individuals in the family. The choice of agricultural produce and their cultivation in Jamaica, as depicted by Brodber, marks the significance of the traditional agricultural practices as opposed to the plantation monoculture within the concerns of the Anthropocene.

When the enslaved people moved away from the monoculture plantation economy and planted multiple crops on their newly owned lands, they broke away from the established patterns of the colonizer. *Nothing's Mat*'s discussion about the plantation in Jamaica revolves around the root staples such as *dasheens*, plantains, and corn. When Brodber depicts the traditional agricultural practices in Jamaica, she illustrates how traditional agricultural practices in the island break away from the preestablished structures of practice and thought. The process of intercropping, the practice of cultivating more than one crop at a time, is a sustainable practice that the community exercised on the provision grounds. The creative method helped maintain soil fertility as opposed to the practice of monoculture followed by the colonizers in their plantations. The novel's unnamed protagonist's ancestor Everard Turnbury, who is the son of Samuel Turnbury and a man of European descent, inherits the ancestor's ancestral land through marriage to the protagonist's great aunt Nothing. The protagonist discusses how Everard uses the land only for banana cultivation and stops the propagation of coconuts and dasheens: "Nobody knew the banana business as well as he. But the *dasheen* and coconuts and so on? They were doing quite well before he came on."[25] The narrative highlights the European interest in monocropping that yields maximum produce returns but depletes the soil quality. Brodber depicts agricultural practices that resist monoculture throughout the narrative, which also discusses the different stages of Jamaican history that resulted in ex-slaves becoming the owners of the land that was once denied to them. Despite Everard Turnbury's disinterest in cultivating dasheen and coconuts, the coconut trees planted on Nothing's birthdays by her father kept propagating. As the coconut trees are almost entirely self-propagated, with one new tree growing wherever its fruit fell, the farm also becomes a coconut estate. The proliferation of the coconuts in the land owned by Nothing and her family across different generations resembles the patterns on her family mat. Nothing also increases the sugar cane cultivation on her land, and the protagonist contemplates adding the sugar cane plant and the coconut which elucidates the significance of perceiving humans and nature as integrated while adopting traditional practices that sustain the land. *Nothing's Mat* reveals that the Jamaican folk practices, traditional agricultural practices, and the close association of the family members with their land are crucial to understanding their islandness in the Anthropocene and to perceiving how microhistories from the Caribbean are essential for islandness to be dissociated from colonial ways of thinking.

Conclusion

Islandness is the "source and reference point for the imagination of Islanders" and informs how we "connect and disconnect parts and wholes" when we observe the interconnection that aspires to transcend historically hierarchical relationships within different communities, species, and their ecosystem.[26] In *Nothing's Mat*, Brodber illustrates this interconnectedness that is integral to islandness through the family mat that ties people to the land and shows how they are not separate or hierarchically above the rest of their ecosystem. When the people depicted in the protagonist's family mat become part of the land and the island itself, the narrative reveals how islandness becomes an alternative way of imagining futures in the Anthropocene where human activities have permanently altered the planet. As Manu Aluli Meyer highlights, "the connectedness of things, gods, people, and land" in island communities is an embedded notion for many island knowledge systems.[27] Islandness, which continues its evolution as a word and emphasizes the significance of interconnectedness among different species and the microhistories from island spaces, becomes important for reimagining land management and sustainable futures in the Anthropocene.

Notes

1 Baldacchino, "Islandness and Prince Edward Island"; Baldacchino, "Interview with Sreya Ann Oommen."
2 Pugh and Chandler. *Anthropocene Islands*.
3 Baldacchino, "Coming of Age of Island Studies," 278
4 Dunn, *Sugar and Slaves*, 151.
5 Dunn, *Sugar and Slaves*, 151.
6 The brochure for the October 1999 conference "Microhistory: Advantages and Limitations for the Study of Early American History," conducted by the University of Connecticut, defined microhistory as

> accounts of hitherto obscure people and events that concentrates on the intensive study of particular lives, communities, and unusual events as prisms for understanding larger cultural and social structures. Microhistories reveal the fundamental experiences and mentalities of ordinary people that broad analyses of nations so often conceal.

See: http://oieahc.wm.edu/conferences/micro/
7 Weale, "Islandness," 81–82.
8 Weale, "Islandness."

9 Baldacchino, *Handbook of Island Studies*, xxi.; McMahon and André, "Literature and the Literary Gaze," 297.
10 Shell, *Islandology*, 2.
11 Baldacchino, "The Coming of Age of Island Studies," 272.
12 Baldacchino, *Handbook of Island Studies*, xxi.
13 Baldacchino, *Handbook of Island Studies*, xxiv.
14 Crutzen, et al., "The Anthropocene," 614.
15 Seo, "Can We Return to Ahupua'a?"
16 DeLoughrey, *Allegories of the Anthropocene*, 2.
17 Mignolo, *The Darker Side of Western Modernity*, 119–120.
18 Mignolo, *Delinking*, 452.
19 Mignolo, et al., "Decolonial Aesthetics (I)."
20 Mignolo, et al., "Decolonial Aesthetics (I)."
21 Brodber, *Nothing's Mat*, 5.
22 Brodber, *Nothing's Mat*, 5.
23 Benítez-Rojo, *The Repeating Island*, 4. My larger research addresses the questions of fractals, repeating islands, and nothingness within the idea of islandness in detail where I examine the idea of fractals in Brodber's *Nothing's Mat* within the context of Antonio Benítez-Rojo's idea of the repeating island and Paul Gilroy's *Black Atlantic*.
24 Muhlfelder, "Soil Futures."
25 Brodber, *Nothing's Mat*, 27–28.
26 Weale, "Islandness," 81–82; Shell, *Islandology*, 2.
27 Meyer, "Native Hawaiian Epistemology," 22–28.

Bibliography

Baldacchino, Godfrey, ed. "The Coming of Age of Island Studies." *Tijdschrift voor economische en sociale geografie* 95, no. 3 (2004): 272.

Baldacchino, Godfrey. Interview by Sreya Ann Oommen. "Islandness and Prince Edward Island." Email, March 6, 2018.

Baldacchino, Godfrey, ed. *The Routledge International Handbook of Island Studies*. Oxfordshire: Routledge and Taylor & Francis Group, 2020, xxi–xxiv.

Benítez-Rojo, Antonio. *The Repeating Island: The Caribbean and the Postmodern Perspective*. Durham, NC: Duke University Press, 2006.

Brodber, Erna. *Nothing's Mat: A Novel*. The University of the West Indies Press, 2014, 27–28.

Crutzen, Paul J., Will Steffen, and John R. McNeill. "The Anthropocene: Are Humans Now Overwhelming the Great Forces of Nature?" *AMBIO: A Journal of the Human Environment* 36, no. 8 (2007): 614–621. https://doi.org/10.1579/0044-7447(2007)36[614:taahno]2.0.co;2.

DeLoughrey, Elizabeth M. *Allegories of the Anthropocene*. Duke University Press, 2019, 2.

Dunn, Richard S. *Sugar and Slaves: The Rise of the Planter Class in the English West Indies, 1624–1713*. University of North Carolina Press, 1972, 151.

Lee, Richard E. "Legitimating Hierarchy and Constructing Consensus, or the 'Cultural' Aspect of the Modern World-System: The Morant Bay Uprising, the Irish Rebellion, and English Franchise Reform." *Review* (Fernand Braudel Center) 30, no. 3 (2007): 215–229. http://www.jstor.org/stable/40241772.

McMahon, Elizabeth, and Bénédicte André. "Literature and the Literary Gaze." In *Routledge International Handbook of Island Studies: A World of Islands*, edited by Godfrey Baldacchino. 1st ed., vol. 1, 297. Routledge, 2020.

Meyer, Manu Aluli. "Native Hawaiian Epistemology: Sites of Empowerment and Resistance." *Equity & Excellence in Education* 31, no. 1 (April 1998): 22–28. https://doi.org/10.1080/1066568980310104.

Mignolo, Walter, et al. *The Darker Side of Western Modernity: Global Futures, Decolonial Options*. Duke University Press, 2011, 119–208.
Mignolo, Walter, et al. "Decolonial Aesthetics (I)." *TDI+Transnational Decolonial Institute*, July 17, 2017.
Mignolo, Walter, et al. "Delinking: The Rhetoric of Modernity, the Logic of Coloniality and the Grammar of De-Coloniality." *Cultural Studies* 21, no. 2–3 (March 1, 2007): 449–514. https://doi.org/10.1080/09502380601162647.
Muhlfelder, Hannah. "Soil Futures: Environmental Management as Care in Mary Ann Shadd Cary's "A Plea for Emigration." Forthcoming manuscript in *Resourcing Love*, edited by Jada Ach and Kristen Brown. New York: Routledge.
Omohundro Institute of Early American History and Culture. 2019. "Microhistories in Early America Conference." Accessed August 11, 2018. http://oieahc.wm.edu/conferences/micro/.
Pugh, Jonathan, and David Chandler. *Anthropocene Islands: Entangled Worlds*. University of Westminster Press, 2021.
Seo, Diane. "Can We Return to Ahupua'a?" *Honolulu Magazine*, March 29, 2025. www.honolulumagazine.com/can-we-return-to-ahupuaa/.
Shell, Marc. *Islandology: Geography, Rhetoric, Politics*. Stanford University Press, 2014, 2.
Weale, David. "Islandness." *Island Journal* 8 (1991): 81–82.

6 Aridity

Gary Reger

In a novel set in the Tuareg Sahara, the narrator encounters a lone traveler in the desert who is

> battling his thirst. He was struggling, with a heroism seen in the desert only among perpetual voyagers who have long familiarity with thirst. Only a person who has firsthand experience of thirst knows that for a thirsty person to resist his desire for water is more heroic than for a cavalryman to charge toward death's portal, because only those with firsthand experience of thirst understand that thirst is death. Indeed, it is a fate worse than death.

The Australian explorer Ernest Giles minced no words: "[I]n a region such as this [the deep desert] we must be prepared to lay down our lives at any moment in our attempts to procure water, and we must take it when we find it at any price." Edward Abbey wrote of "the familiar terrible desert thirst." Crossing the desolate El Houl plain in northwestern Arabia, T.E. Lawrence faced a "breathless wind with . . . furnace taste" whose desiccating heat left "my throat so painful that for three days afterward I could eat little of our stodgy bread."[1] Whence comes this vicious thirst, this "immortal power of the desert"? It comes, most powerfully, from the deserts' *aridity*.

Aridity and its companion thirst present potent challenges to the settler-colonial project of managing western lands of North America (see Figure 6.1)—a challenge that, as we just saw, is also global. But in North America, it became obvious from the first settler-colonial explorations of the vast West that aridity might make European settlement virtually impossible. As early as the Zebulon Pike's and the Stephen H. Long expeditions (1806–1807, 1819–1820), the Great Plains were condemned as the Great American Desert. The phrase was later used repeatedly to describe vast arid stretches of western lands, generally regarded as uninhabitable, hostile, and deadly—because, among other threats, the traveler would die of thirst.[2] Many of these early explorers thought the new territory west of the Mississippi (more or less) would never be occupied, being too dry; indeed, John Wesley Powell, who wrote

DOI: 10.4324/9781003581185-8

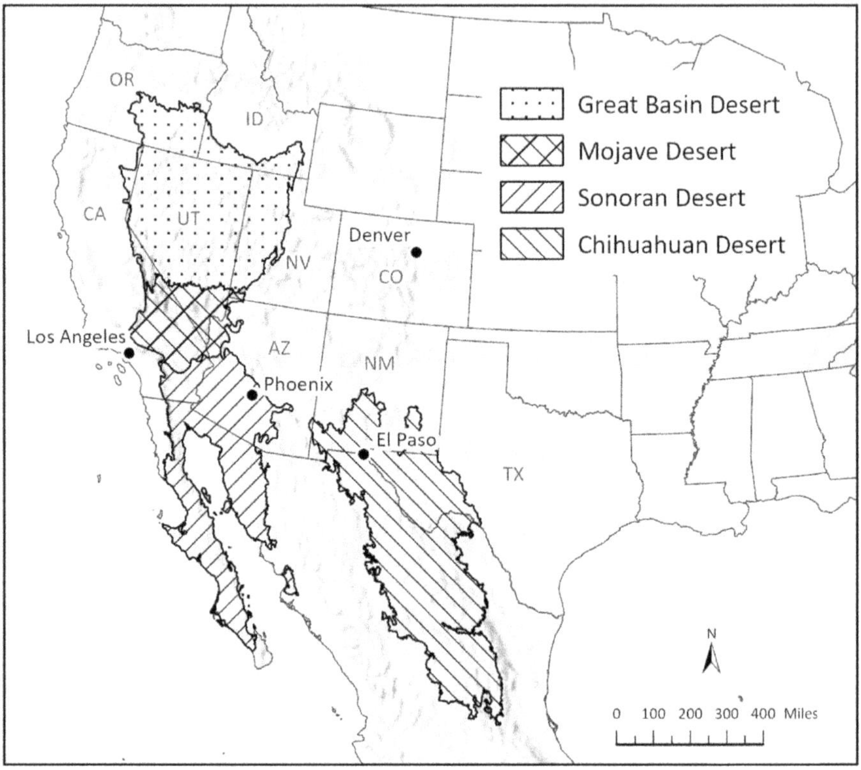

Figure 6.1 Deserts of the American Southwest.

Source: Compiled by Michaele Buenemann, Department of Geography and Environmental Studies, New Mexico State University

the classic study of the arid West, instead of offering a case for settler-colonial management, argued for confining settlement to hydrological regions, in numbers no greater than the water could support. He'd find Phoenix an abomination.

What is Aridity?

In Melissa Broder's recent *Death Valley* (2024), her protagonist asks, "How many times can you use the word *arid*?"[3] Plenty, it seems; aridity is everywhere on Earth. Broder's character doesn't ask the deeper question: what is aridity? The simplest definition follows from low total annual precipitation, the line usually drawn at less than 10 inches (25.4 cm) per year used to mark out the desert zones (B) in the Köppen-Geiger classification of climate zones (see Figure 6.2).[4] A more elaborate definition is the Aridity Index (AI), based on precipitation and evapotranspiration, which is the loss of water to the atmosphere via evaporation.[5]

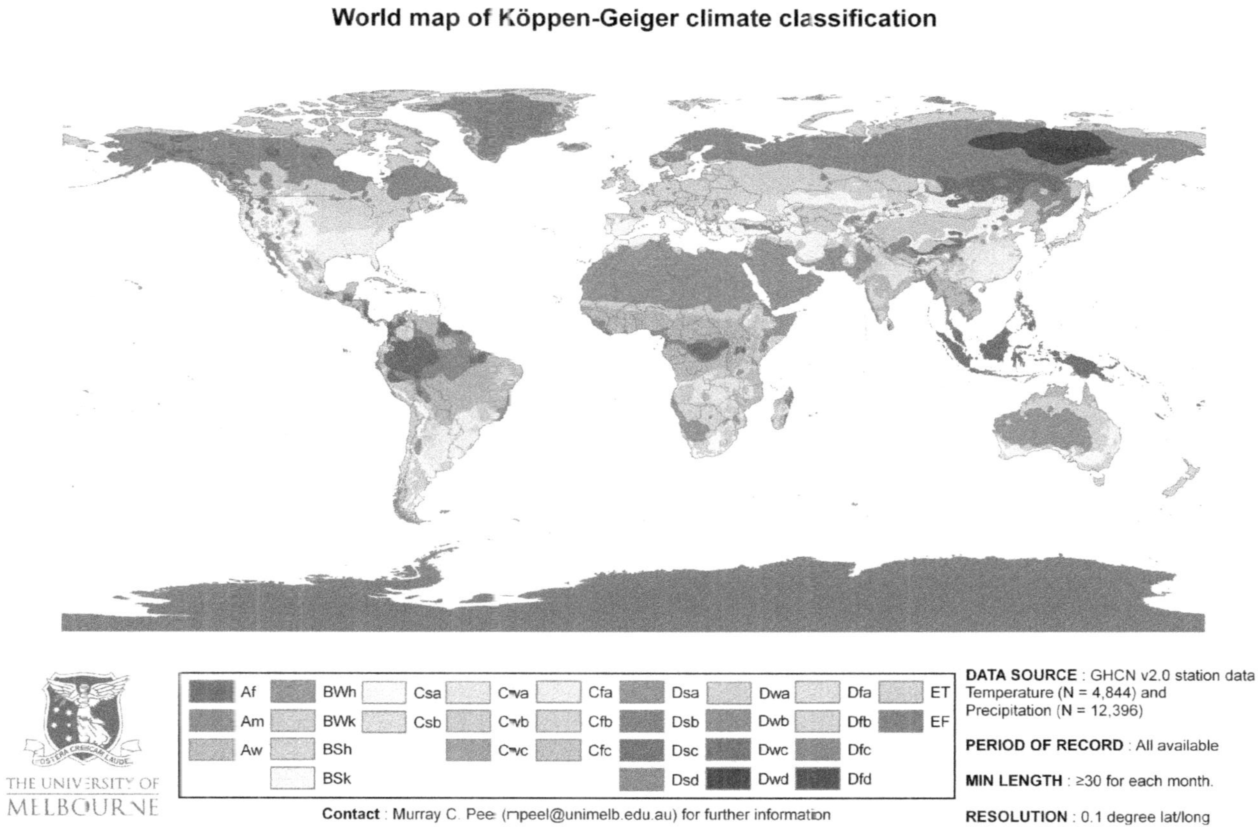

Figure 6.2 The Köppen-Geiger classification of climate zones.

Source: Modified after Peel et al. (2007)

Drivers of aridity include various geophysical and meteorological factors. The bands of arid space that cluster around 30° N and S on the globe are typically a result of the atmospheric circulation pattern called Hadley Cells. Heated and humid, air in the tropics rises to an altitude of roughly 6.2–9.3 miles (10–15 km). As it moves north or south, it is drained of its moisture; at about 30°, it descends, now very dry, heats, and helps create the arid deserts at these latitudes. Another factor can be rain shadows, caused when air masses are driven up high mountains, wrung of their moisture, and then drop on the other side, now dry and hot. The American deserts east of the Sierra Nevada are examples. Cold ocean currents can also create arid deserts by denying evaporation, as in the Atacama in South America and Namibia in southern Africa. Landscapes far from the ocean, like the Taklamakan Desert in Central Asia and much of the Sahara, typically are arid.[6] And ironically, desert air, typically arid, can sometimes be brutally humid: John Annerino, who walked a long stretch of the Camino del Diablo in the Arizonan Sonoran Desert, complained of brutal humidity.[7]

Managing Aridity, Aridity as Manager

Aridity itself cannot be managed; there is no imaginable technology that can overcome Hadley Cells or rain shadows, despite experiments with cloud seeding or the unsupported insistence that "rain follows the plough."[8] To live in an arid environment demands a different sort of managerial accommodation: typically, importing water from elsewhere to irrigate crops, quench parched throats, and more recently, feed golf courses and sustain the growth of urban conglomerations more suitable to the wet East than the arid West—like the abomination of Phoenix. Long before the arrival of Anglo settlers, the Hohokam had diverted the waters of the Salt and Gila Rivers to irrigate their fields. Eventually, though, aridity overwhelmed them, as they probably succumbed to a megadrought that persisted from 1276 to 1299 and, with less intensity, to about 1450.[9] Much later in the Phoenix basin, Jack Swilling, having observed the remains of Hohokam irrigation canals, undertook in 1867 and 1868 their clearing and the construction of new canals, draining water from the Gila and Salt Rivers to support farms that grew produce and hay for nearby military installations.[10] At about the same time, John Wesley Powell released his *Report on the Lands of the Arid Regions of the United States*, in which he argued that the way to settle arid spaces was to respect the limitations imposed by the region's hydrology. That advice, however, was not to be followed. A few decades later in 1902, Congress created a new agency, later named the Bureau of Reclamation, to pock the West with dams, canals, pumps, and massive water-control systems (see Figure 6.3)—an extraordinary managerial effort to crush aridity and transform the dry West into a simulacrum of the wet, fertile East (see the chapter on reclamation by Grover, who deals in more detail with the matters I mention here).

Figure 6.3 All-American Canal (California) seen from inside Hydroelectric Drop Station 4.

Source: Photograph by the author

But this "management" does nothing to alleviate the aridity. If, as new materialists and other contend, environments have agency, then surely aridity—and its companion, heat—serves as a major instrument in desert agency.[11] When you spend time in a western US desert, it doesn't take long to notice the problem of water. The landscape is devoid of running rivers and lakes. Vegetation familiar from more temperate climes is absent; instead, mesquite, cacti and other succulents, lichens, moss, and the invisible world of microorganisms are scattered across the ground (see Figure 6.4). Your body, too, soon feels the effects. If it is hot, you sweat abundantly, but the moisture you exude evaporates so fast you may not even perceive it. You can become dehydrated without realizing it; unless you drink plenty of water, you may begin to feel the impact of heat stress, nausea, headache, irritability. All this is a product of the reigning characteristic of desert spaces: aridity, "the grand fact of dryness."[12]

Arid beyond the Landscape

Aridity goes beyond the environment. In common parlance and advertising, for instance, "arid" accrues multiple, sometimes contradictory, connotations

Figure 6.4 Colorado Desert, California.
Source: Photograph by the author

that can obscure its origins as a descriptor of climate. At times, these uses of arid can overwhelm or disguise its primary link to western deserts. There's Arrid deodorant and antiperspirant, which promises to keep you "Ultra Fresh—You'll feel Ultra Fresh with an uplifting scent of bright citrus, fragrant flowers and soft woods to help fight odor."[13] An "arid" outcome of a project or idea expresses metaphorically failure. "Arid" writing or speaking lacks verve and imagination.[14] Metaphorical "aridity" almost always bears a negative connotation (or denotation) drawn from the contrast in the United States between the wet, productive (in the eastern imaginary), and tamed East and the dreadful stretch of arid lands that impeded Western travel, blocking easy access to the Promised Land of California.

The "management" of aridity, as a means to overcome or eliminate it, clearly, is a fool's errand. Aridity pervades our desert West—and the literature set there and in deserts worldwide—and it is not going away; if anything, with climate change, we face an aridification promising an even drier West beyond the 98th meridian, for climate scientists are now largely agreed that the West is in the grip of yet another megadrought, surpassing even the one that spelled the end of the Hohokam.[15] In my home state of New Mexico, efforts are underway to rewrite water-conservation plans for the more than 30 hydro-regions embedded in the state; arguments have begun about diversion for farming, growing urbanization, ranching, industry, oil and gas extraction, and all the other water-intense activities of a capitalist economy that grew assuming water would always be there. This is yet another management effort, which may yield some results if based not on fantasies about draining aquifers or drawing water from wet mountain states but rather on facing realities about the meaning of aridity. The dry fact of aridity will remain, defining indefinitely, with heat, the character of the American West.

Notes

1 al-Koni, *Anubis*, 164; Abbey, *Desert Solitaire*, 160; Lawrence, *Seven Pillars of Wisdom*, 254; al-Koni, *Gold Dust*, 40.
2 For Pike, see Teague, *The Southwest in American Literature*, 6; for the Long expedition, Bell, *The Journal of John R. Bell*; James, *Account of an Expedition from Pittsburgh to the Rocky Mountains*. See also Baltensperger, "Plains Boomers and the Creation of the Great American Desert Myth"; Sweeney, "Wither the Fruited Plain"; and LeMenager, *Manifest and Other Destinies*, 23–106, among a wide-ranging literature on this topic.
3 Broder, *Death Valley*, 37.
4 Peel and McMahon, "Updated World Map of the Köppen-Geiger Climate Classification," 1633–1644. See https://www.britannica.com/science/Koppen-climate-classification (accessed August 14, 2024) for a good, brief account of the system.
5 United Nations Environmental Program, *World Atlas of Desertification*.
6 Warner, *Desert Meteorology*, 27–53.
7 Annerino, *Desert Survivor*, 16–18.
8 Libecap and Hansen, " 'Rain Follows the Plow' "; Harper, *Make It Rain*.
9 Hill, et al., "The 'Collapse' of Cooperative Hohokam Irrigation in the Lower Salt River Valley."
10 Zarbin, *Two Sides of the River*.
11 See Reger, *Wild, Weird, West*, 193–215.
12 Krutch, *The Desert Year*, 17.
13 https://www.arrid.com/products/ultra-fresh-solid.aspx (accessed September 18, 2024).
14 Examples from the online *Cambridge English Dictionary*, https://dictionary.cambridge.org/us/dictionary/english/arid (accessed September 21, 2024).
15 Fountain, "How Bad Is the Western Drought?."

Bibliography

Abbey, Edward. *Desert Solitaire: A Season in the Wilderness*. Simon and Schuster, 1968.

al-Koni, Ibrahim. *Anubis. A Desert Novel*, translated by William M. Hutchins. The American University in Cairo Press, 2005.

al-Koni, Ibrahim. *Gold Dust*, translated by Elliott Colla. The American University in Cairo Press, 2008.

Annerino, John. *Desert Survivor: An Adventurer's Guide to Exploring the Great American Desert*. Four Walls Eight Windows, 2001.

Baltensperger, Bradley H. "Plains Boomers and the Creation of the Great American Desert Myth." *Journal of Historical Geography* 18 (1992): 59–73.

Bell, John R. *The Journal of Captain John R. Bell, Official Journalist of the Stephen H. Long Expedition to the Rocky Mountains, 1820*, edited by Harlin M. Fuller and Le Roy R. Hafen. Glendale: Clark, 1957.
Broder, Melissa. *Death Valley: A Novel.* Scribner, 2024.
Davis, Diana K. *The Arid Lands: History, Power, Knowledge*. The MIT Press, 2016.
Giles, Ernest. *Australia Twice Traversed: The Romance of Exploration, Being a Narrative Compiled from the Journals of Five Exploring Expeditions into and through Central South Australia, and Western Australia, from 1872 to 1876.* 2 volumes. Sampson, Low, Marston, Searle & Rivington, Limited, 1889.
Harper, Kristine C. *Make It Rain: Control of the Atmosphere in Twentieth-Century America*. University of Chicago Press, 2018.
Hill, J., J. Brett, Patrick D. Lyon, Jeffery J. Clark, and William H. Doelle. "The 'Collapse' of Cooperative Hohokam Irrigation in the Lower Salt River Valley." *Journal of the Southwest* 57 (2015): 609–674.
James, Edwin. *Account of an Expedition from Pittsburgh to the Rocky Mountains, Performed in Years 1819 and '20, by Order of J. C. Calhoun, Secretary of War, under Command of Major Stephen H. Long. From Notes of Major Long, Mr. T. Say, and Other Gentlemen of Exploring Party*. 2 volumes. H.C. Carey & I. Lea, 1923.
Krutch, Joseph Wood. *The Desert Year*. William Sloane Associates, 1952.
Lawrence, T. E. *Seven Pillars of Wisdom: A Triumph*. Penguin Books, 1962.
LeMenager, Stephanie. *Manifest and Other Destinies: Territorial Fictions of the Nineteenth-Century United States*. University of Nebraska Press, 2004.
Libecap, Gary D., and Zeynap Kocabiyik Hansen. " 'Rain Follows the Plow' and Dry-farming Doctrine: The Climate Information Problem and Homestead Failure in the Upper Great Plains, 1890–1925." *Journal of Economic History* 62 (2002): 86–120.
Osuna, Celina, and Aidan Tynan, eds. *Storied Deserts: Reimagining Global Arid Lands*. Routledge, 2024.
Peel, M. C., B. L. Finlayson, and T. A. McMahon. "Updated World Map of the Köppen-Geiger Climate Classification." *Hydrology and Earth System Sciences* 11 (2007): 1633–1644.
Powell, John Wesley. *Report on the Lands of the Arid Region of the United States, with a More Detailed Account of the Lands of Utah*. Harvard Commons Press, 1983.
Reger, Gary. *Wild, Weird, West: Essays on Arid America*. Texas Tech University Press, 2024.
Sweeney, Kevin Z. 2005. "Wither the Fruited Plain: The Long Expedition and the Description of the 'Great American Desert'." *Great Plains Quarterly* 25: 105–117.
Teague, David W. *The Southwest in American Literature and Art: The Rise of a Desert Aesthetic*. University of Arizona Press, 1997.
United Nations Environmental Program. *World Atlas of Desertification*. Edward Arnold, 1992.
Warner, Thomas. *Desert Meteorology*. Cambridge University Press, 2004.
Zarbin, Earl. *Two Sides of the River: Salt River Valley Canals, 1867–1902*. Salt River Project, 1997.

7 (Un)documented Ecologies

Rafael A. Martínez

Saguaros found in Arizona's Southwest borderlands hold a mythical place in the landscape of migrant literature. Salvadoran poet, Javier Zamora, dedicates a passage in *Unaccompanied* (2018), a collection of migration poetry where the author describes the rough terrain in a migrant's journey. In a passage from the poem "Saguaros," Zamora showcases the physical and built presence of the desert landscape while also unmasking an experience felt in the physical body carried as trauma from a migrant episode.[1] Zamora represents a man-made creation of the borderlands through the direct representation of "barbwire" that highlights the militarization of the border, which limits human mobility and the natural world as the line "nothing growing under it" points out. In this case, Zamora is recounting a hallucination experienced in crossing the borderlands as an unaccompanied child migrant and talking to the character of a bat who highlights that "the blood of the saguaro seduces migrants,"[2] pointing to the United States as an enticing destination point. The poem begs the question: migration at what cost? Migrant literature like that of Zamora's highlights the haunting relationship between migrants and built ecologies that the US nation-state produced as a militarized border that has come to represent a history of violence. In its ability to (un)document ecologies, migrant literature captures the violence produced by the nation-state through policy and militarization of the border. The act of (un)documenting ecologies in literature constitutes a dual process endeavor: one of inscription and one of erasure. The inscription entails the literary documentation of state-sanctioned physical violence inflicted upon (im)migrant bodies. Conversely, the erasure involves a counter-discursive process that seeks to heal both corporeal and psychological trauma through narrative strategies that foreground resilience, as embodied in the material and symbolic landscapes of the US/Mexico borderlands. I argue that in the Southwest borderlands, policy and ecology converge as mechanisms of exclusion, where climate-induced hazards and militarized infrastructures disproportionately impact (im)migrant lives. In turn, I define the process of (un)documenting ecologies as a dual process: documenting the violence inflicted on migrant bodies and landscapes, while also undoing that violence through creative expression and alternative world-making.

DOI: 10.4324/9781003581185-9

Why (Un)document Ecologies?

(Un)documenting ecologies is at the intersection of migrant literature and critical ecologies. The Merriam-Webster dictionary defines *ecology* as "a branch of science concerned with the interrelationship of organisms and their environments."[3] Ecologists are preoccupied with understanding the relationship between the natural environment and changes or impacts to said landscapes and relationships. Literature, on the other hand, represents an opportunity to document physical experiences by migrants and the environment.

In the historical and literary sense, a document is an archive or a record that exists about an event or historical experience. In that manner, our bodies produce archival repositories that produce a conscious or, at times, unconscious worldview. In *Undocuments* (2021), literary scholar John-Michael Rivera asks the poignant question, "How do you document the undocumented?"[4] In a sense, Rivera is referring to not only documenting undocumented people, that is, (im)migrant populations lacking legal status or presence in a country but also pointing to the documentation process of that which is often not documented or experienced in a straightforward manner. Rivera goes on to state that the word *undocumented* is a "word that simultaneously documents, inscribes, and erases; it creates a condition somewhere between the living and the dead, a spectral speech act that reveals the anxious feeling of *being* Latinx" (emphasis original).[5] And therefore,

> to document the undocumented, to undocument the documented, haunts Latinx existence; the inquiry bespeaks a spectral Latinx subject rendered by the shadow of the state, cast and collected by a parchment doppelganger that leaves us split between paper and flesh.[6]

The connection between "paper and flesh," as Rivera's work points to, is where I want to move the conversation into an ecological relationship between landscape and archival memory. Convergently, scientists have argued that plants, environments, and nature similarly have the ability to hold on to memories and change accordingly to real-life events.[7] (Un)documenting is the process by which we document not only the real effects of a militarized border ecology and the violence it inflicts on human beings, animals, and environments but also the creative ways in which (im)migrants can imagine alternative world-making possibilities.[8] In this way, my work is in conversation with Mika Kennedy's chapter (Chapter 11) of this collection, whose analysis of the novel, *Treadmill,* serves a refusal of settler-colonial violence that prioritizes a triumphal narrative through the main character who openly resists enacting an "unmanageable life." It is vital to (un)document ecologies as a way to denounce and document environmental, human-rights, and social justice violence produced in the borderlands, as well as capture literary consciousness to show how all of these processes are interrelated.

Managing or (Un)documenting Land/Border Ecologies

Historian Rachel St. John details efforts of drawing the *line in the sand* at the Southwest borderlands as an effort where the US government called on geographers, anthropologists, historians, military, ecologists, and others to use landscape markers that would secure the best natural resources toward a growing Empire.[9] The making of the border required a settler-colonial project of displacing Indigenous peoples (genocide) and eliminating any Hispanic or Mexican influence over the landscape (Anglo Fantasy Heritage) by producing the opportunity for Anglo-American populations to have access to land to produce capital.[10] Colonialism ensures the violent replacement of an Indigenous population with an invading settling population aimed at producing capital extraction out of the land.[11] The Immigration Act of 1924 led to the establishment of the Border Patrol. The colonial project must maintain sovereignty in managing a border; but most importantly, it must maintain capital relationships produced in managing a labor force. (Im)migrants have remained an underclass labor force for capital growth throughout America's global economic imperial expansion.

Policy is used to regulate, exclude, and control (im)migrant populations in moments of heightened xenophobia, as seen in historical examples like the Great Depression (1929–1939) or through violent expulsions like Operation Wetback (1954).[12] The physical management of the Southwest borderlands began to take shape through immigration and economic policies like the Immigration Reform and Control Act (IRCA, 1986) and the North American Free Trade Agreement (NAFTA, 1994) where the idea has been to facilitate the flow of economic goods while restricting human mobility through border security.[13] In the 1990s, "Prevention Through Deterrence" was a policy of border enforcement that would funnel immigrants through the hot and inhospitable Sonoran Desert corridor in Arizona to "deter" migrants from crossing.[14] As Dominique Aurilla Vargas examines in Chapter 14 of this volume, "Prevention Through Deterrence" has led to a growing number of deaths.

(Un)documented Ecologies in Literature

In literary expressions, (un)documenting ecologies is two-fold exercise. It requires contemplating the symbiotic relationship between documenting violence experienced in (im)migrant bodies at the hands of the state and calls on readers to examine imaginary acts of resilience and possibilities represented in the expansive landscape. The migration novel, *The Devil's Highway: A True Story* (2005), by Mexican American poet and novelist, Luis Alberto Urrea, references the Sonoran desert ecologies as the "devil's highway" in which the landscape serves as a natural corridor that is weaponized against migrant crossings.[15] The chapter titled, "The Trees and the Sun," proves to be a painstaking exercise in detailing a landscape of trees as "a hurricane of sunlight, and like storm victims, the men hugged the hot trunks, clutched the trees to keep out the killer sun, even tied themselves to the trees."[16] Migrants "clutch" the trees for the purpose of shade, but really for life, while also producing an archive or record of the ways in which the natural landscape produces physical and deadly violence on real bodies. Literary expressions have the ability to produce an active archive showcasing the brutal ways in which the United States uses the Southwest borderlands ecology as a weapon by intertwining immigration policy vested on the idea of "deterring" migrants from crossing harsh conditions.

Even in near-death conditions, Urrea also (un)documents the desert ecologies as spaces of resilience for migrants when he states that, "they found some prickly pears—tuna in Spanish. . . . [T]he liquid in the cactus fruits spared him."[17] In this passage, Urrea reveals a shifting relationship between land and life shaped by the movement of migrants whose crossings challenge and reconfigure dominant land management regimes; these alternative ecologies captured in migrant testimonials expose how landscapes are not only traversed but also contested, offering survival strategies in life-and-death conditions. Urrea would describe border narrative as "testimony [that] reads like modern poetry," pointing out to readers that the essence of modernity is the complex relationship between violence produced at the cost of sovereignty, capitalism, and colonial relationships.

Despite the overwhelming daily oppressions faced by migrants, migrant literature reveals that (un)documenting ecologies can serve as a form of everyday resistance and a source of possibility. Creative expression found in literature and poetry of *undocupoets* like Yosimar Reyes reminds the (im)migrant community to prioritize centering *UndocuJoy* or everyday forms of experiencing joy in a society where falling victims to state violence is the norm.[18] Scholar, writer, and artist Alan Pelaez Lopez builds from Reyes's concept of "undocujoy" to argue that migration literature encourages readers "to love and mourn in the age of displacement."[19] To (un)document ecologies is an exercise not only in detailing the violence inflected onto migrant bodies to be consumed but also to show alternative possibilities or realities of existence for (im)migrant communities. In this way, migration literature allows us to reimagine land management in the Southwest borderlands by challenging the violent logics of exclusion and control and instead foregrounding migrant relationships to land as sites of care, memory, and resistance.

Notes

1 Zamora. *Unaccompanied*, 11.
2 This is my own translation/interpretation of Javier Zamora's words in Spanish.
3 Merriam-Webster, Ecology
4 Rivera, *Undocuments*, 4.
5 Rivera, *Undocuments*, 4.
6 Rivera, *Undocuments*, 4.
7 See Mancuso, *Brilliant Green*; Johnstone, et al., "Changing Disturbance Regimes," 369–378. Auge, et al., "Plant Environmental Memory," plad032–plad032.

8 Martinez, *Illegalized*, 13.
9 St. John, *Line in the Sand.*
10 For more on the "Anglo Fantasy Heritage" framework, see Otero, *La Calle*; Montgomery, *The Spanish Redemption.*
11 See Wolfe, *Settler Colonialism and the Transformation of Anthropology* and Wolfe, "Settler Colonialism and the Elimination of the Native," 387–409.
12 To read more on the period of the Great Depression and its relationship to xenophobia toward immigrant populations in the United States, see FitzGerald and Cook-Martín, *Culling the Masses*; Hirota, *Expelling the Poor*; García, *Operation Wetback*; and Hernandez, *Migra: A History of the U.S. Border Patrol.*
13 For more on NAFTA's impacts and policy development, see French and Manzanarez, *North American Border Conflicts*; and for IRCA readings, see Baker, *The Cautious Welcome.*
14 To learn more on the political philosophy behind "Prevention Through Deterrence" and its impact on migrant death along the Sonoran Desert, see De León, *The Land of Open Graves* and Dunn, *Blockading the Border and Human Rights.*
15 Urrea, *The Devil's Highway.*
16 Urrea, *The Devil's Highway*, 162.
17 Urrea, *The Devil's Highway*, 166.
18 Reyes, "Share the UndocuJoy!"
19 Pelaez Lopez, *Intergalactic Travels.*

Bibliography

Auge, Gabriela, Valentin Hankofer, Martin Groth, Rea Antoniou-Kourounioti, Irja Ratikainen, and Christian Lampei. "Plant Environmental Memory: Implications, Mechanisms and Opportunities for Plant Scientists and Beyond." *AoB Plants* 15, no. 4 (2023).

Baker, Susan Gonzalez. *The Cautious Welcome: The Legalization Programs of the Immigration Reform and Control Act.* Rand Corp., 1990.

De León, Jason. *The Land of Open Graves: Living and Dying on the Migrant Trail.* University of California Press, 2015.

Dunn, Timothy J. *Blockading the Border and Human Rights: The El Paso Operation That Remade Immigration Enforcement.* 1st ed. University of Texas Press, 2009.

Erickson, Winston P. *Sharing the Desert: The Tohono O'odham in History.* University of Arizona Press, 2021.

FitzGerald, David, and David Cook-Martín. *Culling the Masses: The Democratic Origins of Racist Immigration Policy in the Americas.* Harvard University Press, 2014.

French, Laurence Armand, and Magdaleno Manzanarez. *North American Border Conflicts: Race, Politics, and Ethics.* 1st ed. CRC Press, 2017.

García, Juan Ramón. *Operation Wetback: The Mass Deportation of Mexican Undocumented Workers in 1954.* Greenwood Press, 1980.

Hernandez, Kelly Lytle. *Migra! A History of the U.S. Border Patrol.* University of California Press, 2010.

Hirota, Hidetak. *Expelling the Poor: Atlantic Seaboard States and the Nineteenth-Century Origins of American Immigration Policy.* Oxford University Press, 2017.

Johnstone, Jill F., Craig D. Allen, Jerry F. Franklin, Lee E. Frelich, Brian J. Harvey, Philip E. Higuera, and Michelle C. Mac. "Changing Disturbance Regimes, Ecological Memory, and Forest Resilience." *Frontiers in Ecology and the Environment* 14, no. 7 (2016): 369–378.

Mancuso, Stefano, and Alessandra Viola. *Brilliant Green: The Surprising History and Science of Plant Intelligence.* 1st ed. Island Press, 2015.

Martínez, Rafael A. *Illegalized: Undocumented Youth Movements in the United States*. University of Arizona Press, 2024.
Menchaca, Martha. *Recovering History, Constructing Race*. University of Texas Press, 2010.
Montgomery, Charles. *The Spanish Redemption: Heritage, Power, and Loss on New Mexico's Upper Rio Grande*. University of California Press, 2002.
Otero, Lydia R. *La Calle: Spatial Conflicts and Urban Renewal in a Southwest City*. The University of Arizona Press, 2010.
Pelaez Lopez, Alan. *Intergalactic Travels: Poems from a Fugitive Alien*. The Operating System, 2020.
Perez, Emma. *The Decolonial Imaginary: Writing Chicanas Into History*. Indiana University Press, 1999.
Reyes, Yosimar, "Share the UndocuJoy!" *Define American*, August 22, 2017. https://youtu.be/V1kkdBjASfc.
Rivera, John-Michael. *Undocuments*. The University of Arizona Press, 2021.
St. John, Rachel. *Line in the Sand: A History of the Western U.S.-Mexico Border*. Princeton University Press, 2011.
Torget, Andrew J., and Gerardo Gurza Lavalle, eds. *These Ragged Edges: Histories of Violence along the U.S.-Mexico Border*. The University of North Carolina Press, 2022.
Urrea, Luis Alberto. *The Devil's Highway: A True Story*. Back Bay Books, 2005.
Wolfe, Patrick. "Settler Colonialism and the Elimination of the Native." *Journal of Genocide Research* 8, no. 4 (2006): 387–409.
Wolfe, Patrick. *Settler Colonialism and the Transformation of Anthropology: The Politics and Poetics of an Ethnographic Event*. Cassell, 1999.
Zamora, Javier. *Unaccompanied*. Copper Canyon Press, 2017.

8 Fugitive Sand

Jada Ach

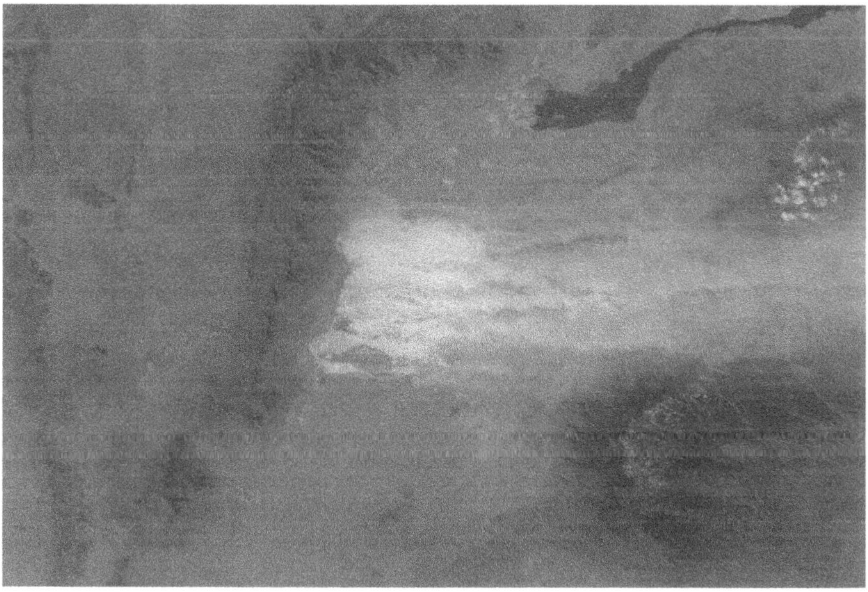

Figure 8.1 White Sands National Monument Dust Storm, New Mexico, USA.
Source: NASA Earth Observatory, 2012

This is a satellite image of New Mexico's White Sands National Park (WSNP) during a wind storm in 2012 (see Figure 8.1). In the region's vast gypsum dunefields, sand is ever on the move, effortlessly disregarding park and military boundaries. We read in the image a story of sand blurring administrative borders—of sand taking flight, feathering eastward like promiscuous smoke. To the satellite, sand reveals itself for what it is: a highly mobile and, therefore, highly unmanageable element. Evading containment and acting at multiple scales—granular, local, global, metaphorical, material, etc.—sand operates as a fugitive material that calls on both land managers and ecocritics to contemplate what it means to live in/with vibrant, wondrous, and often

DOI: 10.4324/9781003581185-10

challenging environments. By sinking our bare feet into the fugitive sand at WSNP, we might learn to understand and, in turn, *love*—and I use this term not only intentionally but also strangely—an eco-material that frequently frustrates and confuses us. In *Emergent Strategy*, writer and activist adrienne marie brown refers to "love experience as data," and she considers it just as important as "any other data" that she uses to guide her community-building work.[1] In this desert-focused keyword entry, I take inspiration from brown's conception of "love data" and use it to reconsider our relationships to the nuclear environments of the American Southwest. From this sandy love, more just ways of managing (with/in) these lands might emerge.

One cannot think about WSNP's vibrant gypsum dunefields without also thinking about the active missile range that surrounds it, the Air Force Base that sits at its eastern border, the federal agency (with War Department roots[2]) that now manages it, as well as the broader context that is New Mexico's atomic history. After all, White Sands Missile Range, which shares a "Zone of Cooperative Use" with the national park, is where the US Government detonated the first atomic bomb at the Trinity Test Site in 1945, a mere 60 miles north of the park's official boundary. To this day, the park periodically closes to visitors to accommodate nearby missile tests, and the sand, no doubt, carries the material traces of those and all previous detonation events. (Errant missiles, unmanned drones, and aircraft have even occasionally landed within park boundaries, and unexploded ordnance exists within the Zone of Cooperative Use.[3]) WSNP is very much a militarized, nuclearized environment despite its designation as "one of the world's great natural wonders."[4] Given these competing meanings and management objectives, it is easy for the park's eco-material complexities to get lost or abstracted. After all, WSNP is very much an intensely managed and surveilled space within a series of other managed and surveilled spaces: a tract within a tract within a tract. Within these related, but distinct, management regimes, NPS tells diverse scientific and economic stories about the white gypsum sand—and just as many methods are used to control it, including grading and hard-packing park roads and insulating septic tanks to prevent clogs. However, the sand communicates its own fugitive stories—stories of flight, encounter, and human engagement[5]—and through "love experience," those stories arrive at our feet in waves of unmanageable delight.

Arthur Sze's 2016 poem "White Sands" buries its body in the fugitive sand and resurfaces as an invitation to drift into more expansive ecological engagements—or, as Sze puts it, to "move as a dune moves."[6] Sze's poem contemplates WSNP as a dynamic environment that actively collaborates with the human, providing readers with sandy insights into how to listen to, experience pleasure in, and, ultimately, survive within this precarious ecological moment. The flecks of white gypsum sand that flutter around and stick to bodies in Sze's poem queerly extend beyond corporeal, temporal, and geographic boundaries; this vibrant elemental unfurling, as well as the pleasure and wonder it inspires in the speaker, demonstrates a radical response to the monolithic story of the desert as a barren wasteland. Sand's spirited response is especially important given the complicated nuclear histories that have contributed to the making of WSNP. Sze envisions the intensely managed dune field as a radical *drift-space* where bodies can rest, be generously unproductive, and relax deeply into the political and ecological possibilities of geo-love.

At first, Sze's "White Sands" appears to communicate a rather romantic vision of human-environmental engagement, one that appears uncritical of or even oblivious to the park's atomic context. The rolling, saltating appearance of the poem feels incredibly dreamy and dune-like, which invites opportunities to explore the curious relationship between poetic form and ecological engagement. The story about how the poem came to be also resonates with the romantic nature-writing tradition in the United States, a tradition that often precludes critiques of settler-colonial policies and practices. "White Sands" was commissioned by the Academy of American Poets (AAP) in 2016 and was funded by a National Endowment for the Arts (NEA) Imagine Your Parks grant, a grant which was meant to "connect people with the National Park System and its programs."[7] The park service celebrated its centennial in 2016, the same year the poem appeared as a standalone on the AAP website. Therefore, it is easy to read "White Sands" as a sentimental ode to one of the country's "special places."[8] However, the poem registers a bit differently—as more of a nuclear poem than a romantic nature poem—when read in the context of Sze's poetry collection, *Sight Lines*, published three years later in 2019. When we reorient our vision to the fugitive nature of sand in Sze's poem, WSNP's nuclear realities begin to materialize; poetry, then, operates as a rich archive for understanding how love has the capacity to inform our approaches to managing land.

What strikes us at first about the poem is its short, two- to three-line stanzas that eagerly roll across the page like gentle waves or dunes. Lines unfold, release their tension, lazily spill their elements and observations, then gather up again for the next offering. This saltating rhythm—spill and gather, spill and gather—remains constant despite the inconstancy of time, location, or subject in the poem. Furthermore, the dashes that start and end the poem tell us that the dunes' queer, unproductive crawl is ongoing. (I use the word "unproductive" because the dune fails to accumulate worth or energy in the capitalist sense.) Time and place drift about in the poem, resulting in a kind of joyful dispersal and disorientation. The poem's sand-energy refuses to be concentrated into a single vessel for atomic deployment; instead, the fugitive energy is expended wildly, joyfully, freely—and without a target.

As we sink down into the sand—or is it the ocean? the sky?—we're unsure if we're operating in the past or the monumental present. At one point, the speaker attempts to stake out a specific location in the timeline by asserting that "an hour earlier, we rolled down a dune." However, the poem then shifts quickly into cactus-time and then yucca-time before leaping, oddly, to "so many years later." Amid all of these crawling turns in time and place, shifts in subjectivity also emerge—"I" and "we" fold in and out of each other, and "your eyelids" morph into "our eyelids." At the same time, Sze repeats the word "as" as a tool that enables human-desert simultaneity. Sentences like "we are lying in a warm dune 'as' a full moon lifts" make room for more-than-human, and even cosmic, co-actors. Time-as-an-*as* is shared—or, to use dune-language, drifts and accretes—generating a sandy logic that management cannot wholly arrest.

With these sandy, dune-like readings in mind, I'd like for us to consider "White Sands" as a nuclear poem—and to explore what that classification might afford ecocritics who are invested in imagining new land management futures. Even though Sze does not refer to the region's nuclear history in "White Sands," his celebration of sand's fugitive, boundary-crossing nature teaches us that loving land that has been radically compromised is fundamental to the process of resisting military-industrial control. As a resident of New Mexico since 1972, Sze is very aware of and troubled by New Mexico's complex nuclear histories and landscapes; he even notes that "From my writing studio, I could see the lights of Los Alamos when it was dark."[9] *Sight Lines* is full of references to these nuclearized landscapes. The collection contains over ten direct references to the state's atomic culture, including descriptions of nuclear bombs and labs, uranium mines, plutonium and germanium waste, shock waves, trinitite glass, and the Trinity Site. Two pages after "White Sands," a chilling line appears on an otherwise blank page: "—The plutonium waste had been hauled to an underground site—" And then a few pages later, in "Talisman," these lines appear: "suddenly/you are walking toward Trinity Site/searching for glass//and counting minutes/of exposure under the sun;/suddenly small things ignite."[10] The Tularosa Basin, it seems, is full of both mineralogical and destructive potential, and Sze repeatedly reminds readers of the dangers that vibrate under the surface of the region's "warm dunes."[11]

Read in these broader contexts, the nostalgic pleasure the speaker expresses in "White Sands" is a slanted kind of pleasure that, like fugitive sand, hides more than it reveals. When the directional movement of the poem finally "lifts"—for the first time—into the "ocean of sky" in the final lines, the now-collective speaker, the "we," remains in the earthly elements. They/we are unable to extricate themselves/ourselves from the fugitive sand, but they/we find love and connection there nonetheless. Instead of being overcome by nuclearization and referencing it, as such, in the poem, the collective, multi-pronouned—speaker in Sze's poem finds something to love in the unstable, radioactive, altogether fugitive sand. The speaker gives sand the space to discover its own logic both within and beyond nuclearity.

In an interview, Sze noted about his writing process that "[i]n letting the poem expand and discover where its terrain is, it goes into a lot of different arenas—and I'm using arenas as equivalent to muses."[12] (I find it interesting that the word "arena" derives from the Latin word for "sand," *harenae.*) Fugitive, arenaceous logic, as it queerly "expands" across the page in "White

Sands," goes underground not to bury (nuclear, settler-colonial) waste or conceal the grave expressions of violence that industrial militarism has wrought, but to seek coolness, radical rest, love, and ecological understanding.

To conclude, "White Sands" offers geologically inspired representations of love and connection in a landscape that is intensely nuclearized, managed, and surveilled. To return to brown, the "love experience" that the poem communicates serves as meaningful data that can inspire more reciprocal management practices and policies at WSNP and other atomic landscapes. The sands produce and accommodate all kinds of life, events, relations, and temporalities. As fugitive matter, sand finds ways to lift and move; it disguises itself as smoke and water and air; and it diversely and continuously frustrates management. At the same time, fugitive sand archives in each grain the love and relations it has absorbed—and the violence it has endured. To know an environment through "moving as a dune moves" is very different from knowing it as a cartographer, land manager, or even missile operator. Moving *as* a dune (there's that "as" again) has the potential to transform White Sands from being a landscape whose regulations would seem to reject wonder, or even life, into a place of pleasurable, granular, unpredictable—*always* unpredictable—becoming.

Notes

1 brown, *Emergent Strategy*, 32.
2 From 1886 to 1918, the United States Cavalry managed national parks "without a legal framework or means of law enforcement." When the NPS took over operations in 1918, much of the military culture remained. Hampton, *How the U.S. Cavalry Saved.*
3 Welsh, *Dunes and Dreams*, 102; National Park Service, "White Sands Incident Reports."
4 National Park Service, "White Sands National Park Official Website."
5 Fossilized human footprints from 23,000 to 21,000 years ago have been discovered within WSNP's boundaries. The sand preserves the story of their presence. Bennett, et al., "Evidence of Humans."
6 Arthur Sze, excerpts from "White Sands" and "Talisman" from *The Glass Constellation: New and Selected Poems*. Copyright © 1976 by Arthur Sze. Reprinted with the permission of The Permissions Company, LLC on behalf of Copper Canyon Press, coppercanyonpress.org.
7 National Park Service, "NEA Imagine Your Parks."
8 U.S. Department of the Interior, "Our National Parks."
9 Sze, quoted in Perry, "Charging the Through Line."
10 Sze, *Sight Lines*, 41.
11 Sze, *Sight Lines*, 36.
12 Sze, quoted in Perry, "Charging the Through Line."

Bibliography

Bennett, Matthew R., David Bustos, Jeffrey S. Pigati, Kathleen B. Springer, Thomas M. Urban, Vance T. Holliday, Sally C. Reynolds, Marcin Budka, Jeffrey S. Honke, Adam M. Hudson, Brendan Fenerty, Clare Connelly, Patrick J. Martinez, Vincent

L. Santucci, and Daniel Odess. "Evidence of Humans in North America during the Last Glacial Maximum." *Science* 373, no. 6562 (2021).

brown, adrienne marie. *Emergent Strategy: Shaping Change, Changing Worlds*. AK Press, 2017.

Hampton, H. Duane. *How the U.S. Cavalry Saved Our National Parks*. Indiana University Press, 1971.

National Aeronautics and Space Administration. "White Sands National Monument Dust Storm, New Mexico, USA." *Photograph*, February 28, 2012. "Images." *NASA Earth Observatory*. https://earthobservatory.nasa.gov/images/77775/white-sands-dust-storm

National Park Service. "NEA Imagine Your Parks." 2016. https://www.nps.gov/subjects/centennial/nea-imagine-your-parks.htm.

National Park Service. "White Sands National Park Official Website." https://www.nps.gov/whsa/index.htm.

Perry, Ayleen. "Charging the through Line: An Interview with Arthur Sze." *Terrain.org*, February 14, 2018.

Sze, Arthur. *Sight Lines*. Copper Canyon Press, 2019.

United States Department of the Interior. "Our National Parks Are National Treasures." Accessed January 20, 2021. https://www.doi.gov/pressreleases/our-national-parks-are-national-treasures.

Welsh, Michael. *Dunes and Dreams: A History of White Sands National Monument: Administrative History*. Santa Fe: National Park Service Division of History, 1995.

9 Rural/Wilderness

Surabhi Balachander

The term "land," especially in the context of public lands in the United States, usually evokes an expanse where the aesthetic heft of the land itself outweighs that of its built structures. Urban public lands tend to be smaller and more obviously engineered—city parks, historic sites, and other spaces carefully built and curated to achieve specific recreational or educational goals. Land outside urban areas (which encompass both cities and their suburbs) is emblematic of US public lands and, in both popular and administrative rhetoric, is either "rural" or "wilderness." Though the terms sometimes blur in popular use, they represent two distinct aesthetic and cultural categories. As I show here, administrative classifications often shape not only popular imagination of these categories but also management practices that differentiate their uses, thus changing the lands themselves. Turning a critical eye to administrative spatiality spotlights the ongoing role of bureaucracy in defining as well as altering lands that most Americans are conditioned to view as "natural."

All land has been shaped by human intervention, but popular rhetoric and bureaucracy construct some lands as more "natural" than others. Rural lands host active cultivation and human habitation, deriving their value at least partly from the production of commodities. Wilderness, on the other hand, derives value from aesthetic beauty, at times ecological significance, and temporary experiences, whether recreational or of the ideal of "preservation." Rural land yields provisioning resources like food and lumber, while wilderness provides more cultural ecosystem services.[1]

As William Cronon notes, the idea of wilderness is itself an urban invention:

> Only people whose relation to the land was already alienated could hold up wilderness as a model for human life in nature, for the romantic ideology of wilderness leaves precisely nowhere for human beings actually to make their living from the land.[2]

In other words, the erasure, both rhetorical and material, of humans creates wilderness. The Wilderness Act of 1964 famously defines wilderness as "untrammeled by man."[3] Yet, "wilderness" has been altered and stewarded

DOI: 10.4324/9781003581185-11

throughout history, primarily by Indigenous peoples. (Additionally, in the age of anthropogenic climate change, no land, with any history, can remain untouched by human intervention.) Histories of settler-colonial removal define both rural and wilderness lands, but in different valences. The federal government seized more agriculturally productive and thus desirable lands for settlers, so while most reservation land today is rural, it is marked by environmental justice issues. Settler-colonial policy forced today's bureaucratic modes of land ownership and cultivation onto tribes and, for example, is responsible for mineral extraction's effects on human health and livestock grazing on the Navajo Nation.[4]

Elsewhere in this volume, Eytan Pol suggests that wilderness might be defined by a (comparative) lack of human control over land. Yet, rhetoric is itself a form of control, and as Pol shows, what occurs in abstract realms—in his case, the commodification of wilderness through trail rating apps—can translate via the administrative realm to material changes. The ideological control of "wilderness" requires both physical and conceptual removal of Indigenous people to reframe lands as "untouched" recreational landscapes for settlers. Rebecca Solnit writes of how, through landscape photography, Yosemite has been "literally pictured . . . as a virgin wilderness," although it "was a battleground before it was a vacation destination."[5] Michael-Shawn Fletcher et al. argue that "wilderness is an inappropriate and dehumanizing construct" with respect to Indigenous peoples,[6] and in *Braiding Sweetgrass,* Robin Wall Kimmerer (Potawatomi) writes, "When we call a place by name it is transformed from wilderness to homeland," suggesting that the term "wilderness" excludes "intimate knowledge" of a place that allows it to be a homeland.[7] In light of settler-colonial dispossession, it is crucial to consider wilderness as an aesthetic and cultural category rather than a quality inherent to certain lands.

"Rural" and "wilderness" are, at their core, also administrative categories. Common definitions of "rural" derive from the United States Census Bureau (anything outside urban areas, in turn, defined by housing unit and population density requirements)[8] and the U.S. Department of Agriculture's (USDA) Economic Research Service (nonmetropolitan counties have some combination of open countryside, rural towns [population <2,500], and urban areas outside of larger metropolitan areas [population <50,000]).[9] Elsewhere, I and others discuss how such definitions, which render rurality administratively spatial, can be limiting and erase rurality's cultural dimensions.[10] In this volume that reckons with the term "land management," however, I am more interested in what the administrative realm *can* do: reshape the aesthetic, cultural, and ecological dimensions of spatial categories like rural and wilderness.

Reading rhetoric from the major federal agencies that manage public lands shows how administrative spatial designations can determine both public perception and material realities of rural or wilderness lands. Most federal public land falls under the purview of two departments (Interior and

Agriculture) and four agencies—the NPS, FWS, BLM, and USFS.[11] Here, I present the missions of these agencies with my own emphases, where bold text denotes wilderness values. Of these, the NPS is probably the best known to the general public, as it manages some of the nation's most iconic (and frequently visited) landscapes: national parks and monuments. Its mission invokes a preservationist and recreation-based wilderness ethic:

> **[P]reserves unimpaired** the natural and cultural resources and values of the National Park System for the **enjoyment, education, and inspiration of this and future generations** . . . cooperates with partners to extend the benefits of natural and cultural resource **conservation and outdoor recreation** throughout this country and the world.[12]

The FWS similarly focuses on conservation, recreation, and cultural values: "working with others to **conserve, protect**, and enhance fish, wildlife, plants, and their habitats for the **continuing benefit** of the **American people**. . . . To **protect** the natural world so **current and future generations** can live with, live from, **and find awe** in lands, waters, and wildlife." The inclusion of "live from" shifts to signal their bridging of conservation with protecting resource production through sites like National Fish Hatcheries.[13] The BLM moves further toward provisioning services, with their mission "to **sustain** the health, diversity, and productivity of public lands for the use and **enjoyment** of **present and future generations.**" Their managed lands host a variety of uses, including "energy development, livestock grazing, recreation, and timber harvesting."[14]

These first three agencies are part of the federal Department of the Interior (DOI), which (as of December 2024) "protects and manages the Nation's natural resources and cultural heritage; provides scientific and other information about those resources; and honors its trust responsibilities or special commitments to American Indians, Alaska Natives, Native Hawaiians, and affiliated Island Communities."[15] This mission foregrounds cultural heritage and, notably, responsibilities to Indigenous groups.[16] The mission does not mention use or production, given DOI's preservationist roots. Instead, "natural resources" are to be protected, and the verb "provide" applies only to information. In contrast, USFS is part of USDA, a department focused far more on administering rural-focused programs and thus on production. USFS works in "providing leadership in the sustainable management, conservation, use, and stewardship of natural and cultural resources on national forests and grasslands in the United States."[17] Its public lands are open to recreational use as well as resource extraction and, crucially, some active cultivation. While each of these federal agencies incorporates some wilderness ethics, reading their missions in conversation with each other reveals that they rhetorically and materially construct public lands on a continuum between wilderness and rural. Each mission invokes futurity in its vision of stewardship, but the terms and goals of this stewardship vary, from preserving recreational opportunities for future generations to ensuring the continued productivity of natural resources.[18]

The aesthetic, cultural, ecological, and administrative realms all intersect in the slippery duality between rurality and wilderness. Each category of land communicates distinct values, and sometimes these values can coexist—or at least, management agencies will them to. Kimmerer, advocating for restoration as an active process of repair, emphasizes the importance of definitional precision: "Restoring land for production of natural resources is not the same as renewal of land as cultural identity. We have to think about what land means."[19] In light of Kimmerer's comments about what land means, I ask an open question: how do we account for administrative spatiality as well as imagine beyond it? Again citing Kimmerer, one possibility:

> Imagination is one of our most powerful tools. What we imagine, we can become. I like to imagine what it would be like if the Honorable Harvest were the law of the land today, as it was in our past. Imagine if a developer, eying open land for a shopping mall, had to ask the goldenrod, the meadowlarks, and the monarch butterflies for permission to take their homeland.[20]

What values can we read into land? Who and what might we imagine as constituents of "rural" or "wilderness" lands?

To close, I ask myself, why do I engage with lands classified—whose classifications even I have internalized—as wilderness or as rural? What do I imagine now about these lands, and what might I need to imagine beyond? Of course, I have my own memories and emotional attachments to these categories that shape my perception of lands. Rurality is front of mind for me because I grew up amidst corn and soy fields in Indiana with a father who worked in agribusiness. Agricultural productivity, often technological, has always contoured what rural means to me. Wilderness is an ideal I absorbed later, as I learned about Western environmentalism in college in California and, alongside it, about the erasure of Indigenous people from supposedly "wild" lands. Though this category is always fraught, there is also something powerful about visiting lands managed as "wilderness," with their exposed geological features and centuries-old trees, landscapes that in their visual appeal and in their obvious history are sublime.[21] When I drove across the country to move from Michigan to Oregon, I recognized on a summer night in Illinois the brand names of seed and agricultural chemical companies planted next to August-tall corn. Later in the trip, I contemplated the obvious geological history of the rocks of the Black Hills in relation to the period that the federal government has controlled them, despite their ongoing, much longer status as a Lakota sacred site. I go to rural landscapes to think about my childhood and to wilderness to think about time.

Yet, to return to this entry's bureaucratic focus, administrative time works on its own scale. My analysis here is a bit of a time capsule, as

I pulled the federal definitions in December 2024, the last full month of Joe Biden's presidency. As I revise, Donald Trump has returned to office, rapidly expanding extraction on public lands. Officials such as Interior Secretary Doug Burgum offer new interpretations of federal definitions that disrupt a "wilderness" ethic: "It says in the mission statement the job of Interior is to 'manage and protect. . . . It doesn't just say 'protect,' it says 'manage and protect.'"[22] Administrative time during this presidency can only be described as accelerationist, threatening sustainable management practices. Administrative spatiality is subject to constant change, and so now more than ever, it is imperative to both understand and imagine beyond its classifications.

Notes

1 Millennium Ecosystem Assessment, *Ecosystems and Human Well-Being.*
2 Cronon, "The Trouble with Wilderness," 17.
3 The Wilderness Act.
4 I am drawing from and thinking with the Diné Policy Institute's report "Land Reform in the Navajo Nation."
5 Solnit, *Savage Dreams*, 221–222.
6 Fletcher, et al., "Indigenous Knowledge," 1.
7 Kimmerer, *Braiding Sweetgrass*, 34.
8 U.S. Census Bureau, "Urban and Rural."
9 USDA Economic Research Service, "Rural Classifications."
10 See: Bell, "The Two-Ness of Rural Life and the Ends of Rural Scholarship"; Bennett, et al., "What Is Rural?" and Balachander, "Rewriting Rural America."
11 U.S. Government Accountability Office, "Managing Federal Lands and Waters."
12 National Park Service, "About Us."
13 U.S. Fish and Wildlife Service, "Mission and Vision."
14 Bureau of Land Management, "Our Mission."
15 U.S. Department of the Interior, "About Interior."
16 Relatedly, from 2021 to 2025, DOI was led by Deb Haaland (Laguna Pueblo), the first Native Cabinet secretary, and NPS by Charles F. Sams III (Cayuse and Walla Walla).
17 U.S. Forest Service, "About the Agency."
18 The emphasis in several of these missions on "future generations" echoes, though does not cite or necessarily reflect, Indigenous values—particularly the Haudenosaunee principle of thinking seven generations ahead.
19 Kimmerer, *Braiding Sweetgrass*, 328.
20 Kimmerer, *Braiding Sweetgrass*, 184.
21 Cronon mentions sublimity as a transatlantic cultural construct central to American conceptions of wilderness.
22 D'Angelo, "Trump to Rescind."

Bibliography

Balachander, Surabhi. "Rewriting Rural America: Race, Environment, and Identity, 1920–2020." PhD diss., University of Michigan, 2024.

Bell, Michael M. "The Two-Ness of Rural Life and the Ends of Rural Scholarship." *Journal of Rural Studies* 23 (2007): 402–415. https://doi.org/10.1016/j.jrurstud.2007.03.003.

Bennett, Kevin J., Tyrone F. Borders, George M. Holmes, Katy Backes Kozhimannil, and Erika Ziller. "What Is Rural? Challenges and Implications of Definitions That Inadequately Encompass Rural People and Places." *Health Affairs* 38, no. 12 (2019): 1985–1992. https://doi.org/10.1377/hlthaff.2019.00910.
Bureau of Land Management. "Our Mission." Accessed December 3, 2024. blm.gov/about/our-\mission.
Cronon, William. "The Trouble with Wilderness: Or, Getting Back to the Wrong Nature." *Environmental History* 1, no. 1 (1996): 7–28. JSTOR.
D'Angelo, Chris. "Trump to Rescind the Public Lands Rule." *High Country News*, April 16, 2025. https://www.hcn.org/articles/trump-to-rescind-the-public-lands-rule.
Diné Policy Institute. "Land Reform in the Navajo Nation." June 2020. dinecollege.edu/wp-content/uploads/2020/06/Land-Reform-In-Navajo-Nation.pdf.
Fletcher, Michael-Shawn, Rebecca Hamilton, Wolfram Dressler, and Lisa Palmer. "Indigenous Knowledge and the Shackles of Wilderness." *PNAS* 118, no. 40 (2021): 1–7. https://doi.org/10.1073/pnas.202221811.
Kimmerer, Robin Wall. *Braiding Sweetgrass*. Milkweed Editions, 2013.
Millennium Ecosystem Assessment. *Ecosystems and Human Well-Being: A Framework for Assessment*. Island Press, 2003.
National Park Service. "About Us." Accessed September 20, 2024. nps.gov/aboutus/index.htm.
U.S. Census Bureau. "Urban and Rural." Accessed January 9, 2023. census.gov/programs-surveys/geography/guidance/geo-areas/urban-rural.html.
USDA Economic Research Service. "Rural Classifications: Overview." Accessed November 29, 2022. ers.usda.gov/topics/rural-economy-population/rural-classifications.
U.S. Department of the Interior. "About Interior." Accessed December 3, 2024. doi.gov/about.
U.S. Fish and Wildlife Service. "Mission and Vision." Accessed December 3, 2024. fws.gov/about/mission-and-vision.
U.S. Forest Service. "About the Agency." Accessed December 3, 2024. fs.usda.gov/about-agency.
U.S. Government Accountability Office. "Managing Federal Lands and Waters." Accessed December 2, 2024. gao.gov/managing-federal-lands-and-waters.
Solnit, Rebecca. *Savage Dreams*. University of California Press, 1994.
The Wilderness Act, Public Law No. 88–577, 16 U.S.C. 1131–1136 (1964).

Part 2

Representations of Management

Histories of Claim and Control

10 Violence against the Land is Violence against the People

Land Management Tactics in Linda Hogan's *Solar Storms*

Jessica Cory

Chickasaw writer Linda Hogan's 1995 novel *Solar Storms* is a fictional account of Indigenous resistance that follows four women living at Adam's Rib, an imagined place inhabiting part of the Northwest Angle at the intersections of lands claimed by Canada and the United States.[1] These women—elder Dora-Rouge, her daughter Agnes, Agnes' biological great-granddaughter Angel/a, and Bush, a Chickasaw friend and early caretaker of Angel/a—travel north from Adam's Rib to defend their homelands, Dora-Rouge's place of birth, from energy conglomerates. The novel's themes include human and beyond-human emplacement, the treatment of humans and beyond-human beings, the effects of damming and water-based violences on places and the beings who inhabit them, and spiritual connections with what we might call the natural world.[2] This chapter not only explores how Hogan draws on histories of damming and Indigenous resistance in her novel but also argues that *Solar Storms* exposes and critiques the myriad ways settler states have utilized "land management" tactics based on "capitalist economics and sustained by settler colonial land policies," as Joshua T. Anderson notes earlier in this collection. These practices/tactics serve dual purposes: freeing up land for state and settler appropriation while continuing the genocidal projects on which such settler states are founded.[3]

The nation-state's construction of hydroelectric dams on Indigenous-held territories is, however, far more truth than fiction. Geoffrey Stacks argues that *Solar Storms* "appears to have been inspired by" the government of Quebec building dams in the James Bay area on Cree territory to divert rivers and harness electrical energy.[4] Known as the James Bay Project, the first phase of this project began in the 1970s, and the second phase began in the mid-1980s. Continual Cree resistance to the project resulted in Hydro-Québec canceling the completion of the project in the early 1990s.[5] As these events would have been ongoing and receiving media attention during the time Hogan was penning *Solar Storms*, Stacks's suggestion that Hogan's novel is based on actual events does hold water. Understanding the construction and impact of state-sponsored infrastructures and their subsequent neglect on these landscapes requires readers to consider the role of political boundaries and the violences infrastructure enacts. For example, erecting physical

DOI: 10.4324/9781003581185-13

structures destroys some ecological boundaries, such as riverbanks that form borders between their waters and drier neighboring areas. In the creation of hydroelectric dams, for example, engineers will frequently reroute rivers in order to meet the demands of the power grid, as Hogan illustrates. Choosing story as a way to reveal these historic and ongoing truths allows Hogan to more easily bring readers into a specific environment and hopefully help them draw connections between what is happening in these northern areas with structural impediments in other places.

However, we must also consider the *place* in which Hogan sets the novel: the Northwest Angle near the Minnesota–Manitoba–Ontario borders. From a literary perspective, locating the novel near national borders highlights their "artificiality," as T. Christine Jespersen observes, as ecological crises do not stop at human-contrived boundaries.[6] Jespersen also suggests that Hogan's choice of location foregrounds the liminal spaces in which Hogan's characters find themselves, a sentiment echoed by Ellen L. Arnold. Hogan describes Adam's Rib as

> the place where water was broken apart by land, land split open by the water so that the maps showed places both bound and, if you knew the way in, boundless. The elders said it was where land and water had joined together in an ancient pact, now broken.[7]

Yet, the permeable boundaries between water and land are just as important as the imposed national borders in Hogan's island setting. The porous spatialities between water and land echo the liminal spaces Hogan's characters inhabit more so than national borders as the former are not artificial. Thus, examining the spaces where water and land interact highlights how water allows for connections between humans and other-than-human entities and pushes back against the colonial water-land binary. Karin Amimoto Ingersoll explains that the land-water binary is a Western idea, historically seen with European travelers who saw journeying across water to be a coast-to-coast mission whereas the Kānaka Maoli view travel as moving *with* water. Epeli Hau'ofa, using terms from his communication with Eric Waddell, explains this false binary as "viewing the Pacific as 'islands in a far sea' and as a 'sea of islands'," arguing that the latter provides "a more holistic perspective in which things are seen in the totality of their relationships."[8] While Hau'ofa is referencing the Pacific and Ingersoll focuses on Hawai'i, their ideas can apply to other coastal or island communities as well, where the liminal spaces created where water meets land offer opportunity for connection rather than division, as Hogan depicts in her novel and as Oommen explains earlier with "islandness." State structures such as dams not only attempt to sever these human-environmental and land-water connections, the structures themselves also work to impose binaries upon stolen lands, binaries that Indigenous and non-Indigenous people, as well as the novel's lands and waters, actively resist.

The construction of hydroelectric dams in Hogan's Anishinaabe Territory began in the early twentieth century. These dams not only impacted the environment due to flooding but also significantly changed lifeways for Anishinaabe families.[9] Brittany Luby notes that "hydroelectric damming between 1898 and 1958 likely contributed to the increased mercury content of family meals," which consisted largely of locally caught fish.[10] Thus, these state violences against the lands and waters negatively affect traditional foods and diets as well as family structures. In a foreword to Brittany Luby's work on the history and impact of damming in Anishinaabe territory, Niisaachewan Anishinaabe Chief Lorraine Cobiness observes:

> Families broke up and left Dalles because of "managed flooding" in the 1960s and 1970s. I can see the lingering damage of dam development on some families from Dalles. Relationships between relatives can be strained because the dams forced them apart for years. Today, when relatives share the same physical space, they sometimes seem to be disconnected.[11]

Chief Cobiness's observations are integral to understanding the full impact and role of these structural impediments. Such structures were built by collaboration between Canadian governments and developers without input from Anishnaabe people, and they are working as designed, in which "the natural boundary of the river is co-opted into the infrastructure" of violence, as Dominique Aurilla Vargas explains in her chapter in this volume.[12] Further, much historical damming has occurred in Anishinaabe territory, where Hogan's fictional Adam's Rib is located, as well as further north as part of broader, state-sponsored "land management" practices. These corporate attempts at land management are state-sponsored violences against the lands and peoples who inhabit them and thus are often met with (particularly Indigenous) resistance. This resistance then, as Joanne Barker, Nick Estes, and other scholars have explored, is often used as an excuse for further state violence.[13]

Since their arrival, settlers have built their colonial-imposed infrastructures upon Native lands now claimed by the United States and Canada, utilizing first the Doctrine of Discovery and later eminent domain to steal Indigenous-held lands (which is to say, all lands) for these purposes.[14] From buildings to railroads to hydroelectric dams, settlers engaged in (and continue to engage in) land-use practices and policymaking that at once lead to water pollution and treat water as a resource or commodity rather than a relative or, at the very least, an agent worthy of respect.[15] Such a lack of respect for water seems somewhat ironic as water travel was key in allowing colonization to occur in the first place. This lack of relationship and related disrespect are due to, borrowing Kevin Bruyneel's term, "settler memory," an often-reproduced political narrative that serves as a method "to legitimate settler governance in relationship to people" by ignoring their modes of governance or positioning such modes of governance as stuck in the past or subordinate to settler frameworks.[16] The political narratives that propel settler memory combine with the limited historical memory of settler presence on lands now claimed by the United States and Canada to further settler political myths of infrastructure as being beneficial to all and as representing/instantiating "progress" and "civilization." Throughout *Solar Storms*, Hogan works to refute these myths, exemplifying Bruyneel's potential for solidarity.

Solar Storms depicts the potential motivation for the corporation's electrification of northern villages and subsequent damming of their waterways for hydroelectric power as what Bruyneel would later theorize as "settler memory." Hogan's characters suggest that these developments are due to not only the lack of settler historical understanding of the peoples and other-than-human creatures who call these northern lands home but also a missing connection to the places they are destroying through such "modernization." Angel, Dora-Rouge's great-granddaughter whose estranged biological mother lives in the northern region, observes at the town's meeting with the developers, "To the white men who were new here, we were people who had no history, who lived surrounded by what they saw as nothingness. Their history had been emptied of us, and along with us, of truth."[17] Notably, Angel's critique of those lacking understanding of these northern landscapes and peoples is geared toward "the white men *who were new here*" (emphasis mine).

This notion of *new* white men forcing "progress" on these northern places and people, however, is not linear or inevitable. Some scholars, such as Christa Grewe-Volpp, argue that Hogan seemingly positions her Native characters against the fallacious "progress" of white settlers. Yet, Hogan's delineation of *new* "white men" implies that white men have already settled in these northern areas and likely have become part of the community, probably following the leads of the region's Indigenous residents in the region, such as Mr. Orensen, who owns the Two-Town Post. Hogan also creates similar communities of people with diverse backgrounds in and near Adam's Rib. However, this idea of historical inaccuracies and erasures combined with positive settler constructions of "progress" allow these new white men to imagine that their work is benefitting the local community by bringing electricity to the northern areas regardless of the cost to the lands, waters, and the area's many residents. As Angel notes, they view the landscape "as nothingness," another *terra nullius*, land that belongs to no one and is ripe for development or white settlement. Framing the contractors' view this way crystalizes the connections between settler colonialism and these later violences against Indigenous sovereignty.

Such structural projects, however, have deeper roots than uninformed white settler contractors. As Bruyneel explains, the impetus for continuing colonization is not simply erasing historical information and continuing to omit accurate details and context but is an extension of it that allows for a deliberate disavowal of Indigenous political configurations and sovereignty. Hogan's non-native contractors are unable to understand and appreciate the connections that the northern peoples have with their homelands, yet the contractors themselves, perhaps unknowingly, enact ongoing projects of colonization through continued land management practices that the northern residents neither desire nor consent to, which completely contradicts the Free, Prior, and Informed Consent (FPIC) required by the United Nations Declaration on the Rights of Indigenous Peoples (UNDRIP). Thus, the contractors refuse to recognize the land as belonging to the Indigenous northern inhabitants and as their responsibility to manage, given their place-based knowledge and multigenerational inhabitation of and relationship with the land. As Auntie, a relative of Angel and Dora-Rouge, explains to the unnamed men in charge of the project, "We've been here for thousands of years. . . . We don't want your dams."[18] Auntie's words and actions in this scene, actions taken to protect the waters and lands she calls home, resist these attacks against her community's land sovereignty. First, she informs the men that her people have "been here for thousands of years," far longer than any settlers, even those few settlers who have become part of the community. Despite Hogan avoiding outright discussions of legal property, this dating emplaces the people in this region, suggesting that because of the length of time they have resided in the region and the relationships they have built with the beings in their homeland, the land management choices should be theirs to make. By privileging Indigenous emplacement and omitting legal claims to land, Hogan demonstrates that to discuss legality regarding property is to validate settler-colonial notions of land and landedness.[19] This lengthy emplacement also suggests that Auntie and her community have place-based knowledge that the corporation's charts and graphs cannot replicate. The original inhabitants' place-based knowledges, part of the "truth" Angel notes earlier, can only be gained through living with other beings long term in a region, through moving with the land's cyclical changes and learning the rhythms that beat like waves against its shorelines.

Auntie speaks from these knowledges while again pushing back against the settlers' refusal to recognize her people's land sovereignty when she tells them, "You've already built a road across the spawning grounds of the whitefish. They'll die from that road. You did it without our permission."[20] First, I want to highlight Hogan's use of pronouns in this scene. Auntie's use of "you" toward the contractors implicates them for both the environmental destruction (building the road) and disrespectful behavior toward Auntie's community (building the road without permission). By using "you" in this way, she is holding them accountable for these acts. Her use of "our" preceding permission is also poignant, as it functions in a possessive yet collective way, illustrating how the land belongs to Auntie's community, as they are in relationship with it, not the contractors. Hogan highlights this relationality through Auntie's knowledge of the whitefish and their spawning grounds, information she assumes the white corporate men do not have and information that one can only receive from living in connection to this place and its waters. By observing the whitefish season after season and viewing the creatures as relatives with needs rather than simply viewing them as expendable without considering the impacts to the greater ecosystem impart place-based knowledge to Auntie. This relationship between Auntie and the fish embodies the definition of "relationality" that Lisa Fink and Lydia Cagluaq Agnus provide here. Again, this knowledge emphasizes a sort of stewardship of these lands and waters; however, such landedness is built on responsibility and reciprocity, a kinship, really, which Auntie invokes by the fact that these men should have sought permission, as the Indigenous peoples who live here are in a reciprocal relationship of belonging to this place.

Chickasaw scholar Jodi Byrd links these types of Indigenous emplacements to sovereignty that necessarily rejects the "rationalizing structures of . . . property ownership within the logics of settler colonialism."[21] Put another way, Auntie's assertion that the men should have sought permission from the land's original inhabitants reinforces Indigenous land sovereignty and self-determination while making clear that that this community does not uphold the ideas of *terra nullius* and the colonial notions of property ownership that follow it, which, in land management terms, require viewing the land as a resource to be exploited. On this latter note, Goenpul scholar Aileen Moreton-Robinson observes owning property as an imperative of white possession or the European and Euro-American need to constantly obtain rather than to belong. While Moreton-Robinson is writing from an Aboriginal Australian perspective, the differences she observes in Indigenous and European (particularly British, but we can extrapolate these concepts to apply to various colonial powers) ideas of selfhood, community, and belonging seem applicable to many Indigenous peoples and nations, including those represented in Hogan's work. By examining *Solar Storms* through these lenses, Auntie's and the other community members' rejection of the dam, electricity, and outside presence emerges not merely as resistance to encroaching and exploitative land management practices but also as an expression of Indigenous sovereignty and self-determination in understanding what is best for one's community and all the beings who belong to that place.

While Moreton-Robinson's understanding of property may seem to offer an Indigenous versus colonizing perspective of land management, Hogan resists the potential binary of settler-colonial/capitalist modes of land management versus Indigenous/decolonial modes. In the novel, she complicates the narrative by including Indigenous characters belonging to a variety of Nations yet inhabiting the same place; integrating settlers into a largely Indigenous northern community; and creating Native characters, such as LaRue Marks Time, who defy the essentialist stereotype of the "ecological Indian."[22] Hogan's novel interrogates ideas of "land management" by revealing how the state's damming projects manage not only physical locations but also the physical bodies constituting the settler state's "Indian problem."[23] *Solar Storms* is set in the early 1970s, which is pertinent for several reasons, including the start of the James Bay Project mentioned earlier and immediately following the era of what has come to be called the Sixties Scoop, in which thousands of Indigenous children in Canada were taken from their families of origin and raised or adopted by white families, as is the case with Angel, Hogan's protagonist.[24]

Early in the novel, Hogan describes the persons living at Adam's Rib—that fictional, liminal space near the Northwest Angle. She writes,

> The first women at Adam's Rib had called themselves the Abandoned Ones. Born of the fur trade, they were an ill-sorted group. Some had Cree ancestors, some were Anishnabe, a few came from the Fat-Eaters farther north. Bush . . . was a Chickasaw from Oklahoma. Others were from the white world; these, the white people, hadn't cared enough for their own kind to stay on with them.[25]

Further down that same page, we learn that LaRue Marks Time, who lives on another part of the island chain, is "mixed-blood from the South, a Dakota."[26] The variety of ancestry among the residents of Adam's Rib and surrounding islands speaks to how emplacement rather than settler notions of biology inform kinship. Some residents are biologically related, such as Dora-Rouge, Angel, and Agnes, yet they are originally from the Land of the Fat-Eaters farther north. Their residence in Adam's Rib, however, highlights the mobility of Indigenous peoples, which often involved waterways, and the ways that many Indigenous peoples interacted and formed communities through trade, marriage, and government force, such as when the state assigned multiple peoples to a single reservation.[27] Thus, their shared emplacement as the foundation for their kinship in turn influences their land management practices, reinforcing the collectivism seen earlier through Angel and Auntie's explanations at the meeting called to discuss the electrification and damming of northern territory.

In order to show how land management methods can shift from being exploitative to reciprocal and respectful, Hogan creates the character LaRue Marks Time, the Dakota gentleman and Vietnam veteran who lives at Old Fish Hook, as an embodiment of this possible transformation. Hogan describes Old Fish Hook as "a nearby settlement on another finger of land that curved like a hook into water and pointed accusingly at Adam's Rib, as if it had sinned."[28] Just as his home is described as misogynistic through its finger-pointing and metaphorically dangerous for nonhuman beings with its barbed shape, Hogan depicts LaRue as having little respect for women and more-than-human beings.[29] He pays Bush to assemble animal skeletons, which he sells to schools and museums. The remains he does not sell, he refers to as "curiosities": "bear teeth, a pheasant with a red face . . . a stuffed bobcat with a cigar in its mouth."[30] LaRue's disrespect of his animal brethren, however, is not limited to those no longer embodied.

In depicting his relationship to humans, fish, and water, when LaRue takes Angel fishing, he refuses to kill the fish he catches, disallowing them a humane death.[31] In fact, he drags the fish alongside the boat and proceeds to skin the creatures while they are still alive, arguing that "they don't have nervous systems."[32] When Angel repeatedly protests and insists he humanely dispatch the fish, LaRue accuses her of "hav[ing] a Bambi complex or something."[33] His dismissal of her concerns and particularly painting her as overemotional and mocking her compassion are clearly chauvinistic and based on his internalized notions of patriarchal superiority. Further, his refusal of humane treatment toward the fish indicates his own repressed trauma that begets violence and limits relationality.

Yet, his internalized machismo counters his ability to achieve the masculine ideal of being an expert fisherman or a provider for his family. Angel observes that LaRue's actions "offend[] the spirits of fish," an observation Bush expands upon when Angel tells her of their fishing trip: " 'There are proper ways of approaching animals and fish. . . . Just as there are proper ways to approach a woman. . . . LaRue knows neither of these."[34] While Bush's remark does suggest a certain heteronormativity, she also makes it clear that animals, fish, and women are similar in their exploitability by LaRue and men like him, though she does add that these entities have diverse spirits and suggests that Angel should approach them differently, as one would approach kin in certain ways.

There is also a tension in this scene between masculinity and indigeneity. While Angel certainly finds LaRue's behavior repugnant, she finds it more problematic that he is Native and yet displays such disregard for the fish and other life forms. She points out that while he talks about how the fish can sense him, he simultaneously has no respect for their senses or intelligence and thus is "a poor excuse for an Indian."[35] The idea of equating a relationship with nature, specifically water and the creatures who live within it, as a mark of "Indian-ness" risks playing into the ecological Indian stereotype. However, Angel seems to understand this (dis)connection between indigeneity and kinship with other-than-human beings as illustrative of how LaRue operates outside of Indigenous considerations of community and what it means to be good kin. Thus, how he manages the land and what he views as its resources are indicative to Angel of his inability to live in community as good kin.

LaRue's ideas of masculinity and indigeneity as related to kinship, however, allow readers to interrogate the role of care and caregiving in social reproduction from a feminist perspective, which, in Hogan's novel, is often associated with land-based relations. The book's main characters (and caregivers) are women: Angel, Bush, Agnes, and Dora-Rouge, though there are male characters in the text who also provide care, namely John Husk and Tommy Grove. We have established that LaRue is misogynistic and disrespectful to most creatures, living or dead (though these binaries are certainly worth interrogating), and that his acts of aggression are due to his own trauma and represent a form of lateral violence. While he attempts to show care for Bush, for instance by providing her with work opportunities and bringing her old maps, his desired outcome is self-serving, as Angel observes that LaRue wants "a woman, any woman. It was his loneliness he wanted relieved."[36] This selfishness-as-motivation and failure to understand his role in the ecosystem and his responsibility to that ecosystem impossibilize not only any potential relationship with Bush but also any potential relationship with his broader environment, including the waters that wind up changing his course.

As a result of the dams that combine the Se Nay and Big Arm rivers, the rivers become, according to Dora-Rouge, "angry."[37] Yet, the water's fury only encourages Dora-Rouge to "talk to the churning river, the white and muddy foam of it, the hydrogen and oxygen of it, and convince it to let us pass safely."[38] After Dora-Rouge's conversation with the water, she reports to her traveling companions, "It will let us go," before offering the river a prayer and tobacco out of respect and gratitude, a recognition that her existence and the river's are intertwined.[39] Her engagement with this water is respectful, and perhaps that is why her experience with it is less fraught than what La Rue experiences.

As a result of the flooding caused by the dam's merging of the rivers, islands in the waterways begin to fracture. As part of the land breaks off near Two-Town, "LaRue leaped on it, as if to pull it back or save it."[40] While LaRue does choose to join himself with the island, perhaps in a masculine act of "saving" the piece of land, this image is less a voluntary baptism and more of an accidental whisking away, as though his actions and the aftermath did not go according to his plan whatever that may have been. He later reappears downriver once Bush and Angel have returned to Adam's Rib, though this span of time is both unknown, and there is no account of what occurs to or with LaRue during the duration he is gone.

Once Angel returns to Fur Island, she is shocked to find LaRue and *The Raven* (his boat) eager to assist her and Bush as they save what they can before Bush's home and the rest of Fur Island sink beneath the floodwaters. Known throughout most of the story for his exploitation and degradation of animals and their spirits, LaRue makes a special effort to tow the land of spiders, a literal island next to Bush's home on which only spiders live, to safety. After witnessing his care for the spiders, Angel believes that his experience on the floating island has changed the man for the better, a representation that audiences can read as the possibility of change for the contractors and their corporations as well. Put another way, if someone as misogynistic and disconnected from the relationality in his ecosystem can change, the developers and governments responsible for the dam and its effects might be able to shift their worldviews as well.

Later, when Angel visits LaRue, she finds him sobbing over the body of an indescribable animal. LaRue explains that a man brought him the creature and claimed the animal was the last of its kind, yet the hunter still killed the species into extinction. LaRue surprises Angel by engaging in any sort of sentiment, as the old LaRue would have considered the creature a mere curiosity, taxidermized it, and put it on display. Hogan tethers LaRue's emotion for the animal to the continued assault on the land and his own unresolved trauma from "the war he'd endured and never told about."[41] While LaRue seems to be referring to prior military experience, we might also view this "war" and his connection with the now-extinct creature as stemming from his own experiences of indigeneity and the violent efforts made by the state to drive Native peoples into extinction. This attempted genocide, of course, was part of a larger land management plan to free up land for settler use.

To further illustrate LaRue's shift in perspective and thus the possibility for a similar change to occur in settler-colonial land management practices, Hogan portrays him in the aftermath of the island's carrying him away. While readers realize that LaRue has experienced a significant metamorphosis near the end of the novel, Hogan's description of this shift is pertinent: "Tears have a purpose. They are what we carry of ocean, and perhaps we must become sea, give ourselves to it, if we are to be transformed."[42] In Hogan's description of tears, the ocean requires humans to "give [them]selves" to receive transformation; there is no alternative option if one desires such change. While Hogan may be using "give ourselves" as a metaphor, meaning to give ourselves over to our tears, our urges to cry, readers might also think about "giving" oneself as a way or building a relationship with the lands, waters, and nonhuman beings who inhabit them. Such a relationship would then necessarily alter land practices to be more just, balanced, and collective and can open up space for transspecies and cross-cultural connection, including the sharing of knowledge and knowledge-building practices, as we see Hogan depict in her novel.[43]

Returning to the scene in which Angel observes LaRue's transformation, Hogan suggests that the ocean is not the only life form with which LaRue shares himself. In this scene, Angel visits LaRue and asks him, "Say, whatever happened on that river when you escaped?" curious to know what events might be behind his transformation.[44] LaRue immediately becomes a bit more guarded, replying, "You think I did that on purpose?. . . I was scared shitless. I thought, This [*sic*] is it. You saw that river. It was wild," obscuring an answer to Angel's question.[45] Quickly following this evasive response, he sees a wolverine from his window, but the creature is only his reflection. LaRue's seeing himself as trickster Wolverine in this now-uncharacteristic act of guarding his feelings further evidences human animality and highlights how other-than-human life forms interact with and transform human lives.[46] When we consider that human/nonhuman and animate/inanimate binaries are colonial constructs alongside how water blurs these boundaries, we can begin to see how relational land management practices offer opportunities for connection beyond settler limitations.

Through juxtaposing pre-dam life in Adam's Rib with the destruction caused by damming the northern territories and combining the Se Ney and Big Arm rivers, Hogan is not only providing a warning against settler-colonial land management practices but also highlighting the importance of intergenerational Indigenous feminist resistance in offering insights and alternative ways of being of the land. Further, Hogan's depictions of LaRue offer a metaphor through which readers see such change as possible. The solutions and alternatives Hogan offers are steeped in both place-based knowledge and community care.

Notes

1 The Northwest Angle is in the Lake of the Woods area of northern Minnesota where Minnesota, Manitoba, and Ontario meet. The Northwest Angle is claimed by the United States due to an erroneous understanding of maps bearing the location of the Mississippi's headwaters (see Lass).

2 I define "emplacement" as being in or of a place, which is not as easy as simply calling a place home and depends on many factors including whether one's ancestors have resided there for millennia or relocated forcibly or by choice.

3 While *Solar Storms'* setting suggests that the dams affected rivers in northern areas and the James Bay Project directly threatened waterways in the northeast (those impacts, of course, affect downstream environs as well), these practices have been widely used in many areas claimed by the United States. For example, Holly Jean Richard and Paul Formisano discuss damming out West and how such practices are depicted in McNickle's, *Wind from an Enemy Sky*; Marmon Silko's, *Gardens in the Dunes*, and Cook-Lynn's, *Aurelia*. Michael Lawson also explores the impact of the Pick-Sloan Flood Control Act of 1944 on Indigenous peoples in relationship with the Missouri River in his *Dammed Indians* and its follow-up *Dammed Indians Revisited*.

4 Stacks, "A Defiant Cartography," 161–162.

5 For more general information on the James Bay Project, see McCutcheon, *Electric Rivers*. For information on the ecological effects of the project, see Hornig, *Social and Environmental Impacts of the James Bay Hydroelectric Project*; Royer, *Climate, Environment and Cree Observations*.

6 Jesperson, "Unmapping Adventure," 294.

7 Hogan, *Solar Storms*, 21.

8 Hau'ofa, *We Are the Ocean*, 31.

9 For a broader Indigenous-focused environmental history of the Great Lakes region and the environmental impacts of colonization, see Hele. For Indigenous responses to Great Lakes area environmental concerns, see Bellfy, *Honor the Earth*; Whyte, et al., "Supporting Tribal Climate Change"; and Whyte, et al., "Seven Indigenous Principles." For Lake-of-the-Woods-specific Indigenous resistance and sovereignty efforts, see Maxwell, "Ojibwe Activism, Harm Reduction and Healing in 1970's Kenora, Ontario"; Luby, *Dammed*; and Willow, *Strong Hearts, Native Lands*.

10 Luby, *Dammed*, 5.

11 Luby, *Dammed*, xiii.

12 While Vargas is describing the Rio Grande river and how the Texas Department of Public Safety has installed buoys with "serrated metal plates," webbing, and razor wire along the riverbanks, readers can see how these governmental changes to rivers, particularly rivers near borderlands, become co-opted into infrastructures that deliberately harm (often non-white) people.

13 To learn more about Indigenous studies scholarship on dams and damming, see Vine Deloria Jr.'s foreword in Lawson's *Dammed Indians Revisited*, in which Deloria argues, "The Pick-Sloan Plan was, without a doubt, the single most destructive act ever perpetrated on any tribe by the United States" (xv); also see Luby's *Dammed* and Bilharz's *The Allegany Senecas and Kinzua Dam*. While these

sources focus on dams built on lands claimed by Canada and the United States, the construction of dams affects Indigenous peoples globally as well, including in China, Brazil, and many other places. For information on Indigenous activism, see Estes and Dhillon, *Standing with Standing Rock*; Spice, "Fighting Invasive Infrastructures."

14 Indigenous peoples residing on lands currently claimed by the US and Canada already had different types of infrastructure such as community buildings and trade routes. For more on Indigenous infrastructures and how TEK might impact contemporary infrastructure and engineering, see Toso, " 'Keeping the Road Clear between Us'."

15 On a similar note, c explores DAPL and the Colorado River Compact in her chapter on Silko in this volume. The concept of agential entities (including nonhuman and non-animal creatures such as water) is a key tenet of object-oriented ontology (OOO), a philosophy that influences my own environmental understanding. For environmentally focused object-oriented ontology, see Morton, especially *Ecology without Nature* and *Hyperobjects*. OOO is also linked to material ecocriticism; see Alaimo. Further, Indigenous scholars like Zoe Todd in "An Indigenous Feminist's Take on the Ontological Turn" have criticized these academic theories for often neglecting attribution to Indigenous worldviews that already held these tenets (i.e., sentience of nonhuman actors).

16 Bruyneel, *Settler Memory*, 16.

17 Hogan, *Solar Storms*, 280.

18 Hogan, *Solar Storms*, 280.

19 The best explanation of "landedness" I have seen comes from Vicente M. Diaz and J. Kehaulani Kauanui who draw from earlier theories of place by James Clifford and Stuart Hall. They build on Clifford's definition of landedness as a "grounding" which " 'offers a sense of depth and continuity running through all the ruptures and attachments' of neo- and postcolonialism, of diaspora and exile" (331). This perspective differs from settler-colonial ideas of land and being in place that are frequently tied to notions of property ownership. "Landedness" in my usage differs from the earlier "emplacement" in that the latter is not necessarily limited to a physical, three-dimensional locale.

20 Hogan, *Solar Storms*, 281.

21 Byrd, "What's Normative Got to Do with It?," 2.

22 For more on this problematic trope, see Krech, *The Ecological Indian*.

23 These "solutions" were used all across North America beginning much earlier than Hogan's novel. Jada Ach notes of Sarah Winnemucca's *Life Among the Piutes*, penned in the latter part of the nineteenth century, that Winnemucca describes how these "built artifacts in the arid and semiarid bioregions of the West . . . impact the social, cultural, and biological lives of humans and nonhumans alike" (129).

24 To learn more about this period, its policies, and their impact, see Cardinal's *Ohpikiihaakan-ohpihmeh (Raised somewhere else)* and Stevenson's *Intimate Integration*.

25 Hogan, *Solar Storms*, 28.

26 Hogan, *Solar Storms*, 28.

27 When designating land as reservations for Native Peoples, the US federal government frequently emplaced multiple peoples, often those who were living near one another, especially if small in number from invasion and genocide, on the same reservations. In many cases, these reservations house multiple peoples, though there is, of course, overlap, as tribal citizens marry or become kin in other ways. Contemporary examples include the Shoshone and Bannock who share a reservation following the Fort Bridger Treaty of 1868; however, as with many reservation lands, theirs were significantly decreased by allotment and increased white settlement.

28 Hogan, *Solar Storms*, 28–29.
29 This correlation between the mistreatment of women or feminized others and the mistreatment of more-than-human creatures (and the Earth more generally) is the main tenet of a branch of ecocriticism known as ecofeminism, a philosophy that gained footing in the early 1990s, though it was described decades earlier. Ecofeminism is based on the concept that the systems of oppression working against women, feminized others, and more-than-human creatures also oppress the Earth, environmental sustainability, and non-white peoples. For more information on ecofeminism, see Mies and Shiva, Warren, and Plumwood. For discussion of ecofeminism in *Solar Storms*, see Schultermandl, "Fighting for the Mother/Land."
30 Hogan, *Solar Storms*, 95.
31 Zoe Todd's work on human-fish relations, "Refracting the State Through Human-Fish Relations," is helpful in understanding some of the underpinnings of LaRue's actions and why they are so problematic.
32 Hogan, *Solar Storms*, 83.
33 Hogan, *Solar Storms*, 83.
34 Hogan, *Solar Storms*, 84.
35 Hogan, *Solar Storms*, 83.
36 Hogan, *Solar Storms*, 131.
37 Hogan, *Solar Storms*, 193. Also, in an interesting connection of Indigenous knowledge, Mercy Oduyoye, a Ghanaian theologian, uses phrasing almost identical to Dora-Rouge when describing her traditions in an interview with Melanie L. Harris:

> In my tradition when there is a flood, a river overflows, we always say the river is angry, the river is angry and when you look you will find that they [people not of that community] have blocked the river someplace with rubbish that people have thrown in or something.
>
> (Harris, "Whiteness as Property," 104)

38 Hogan, *Solar Storms*, 193.
39 Hogan, *Solar Storms*, 193.
40 Hogan, *Solar Storms*, 326.
41 Hogan, *Solar Storms*, 339.
42 Hogan, *Solar Storms*, 340.
43 Hogan, *Solar Storms*, 85.
44 Hogan, *Solar Storms*, 349.
45 Hogan, *Solar Storms*, 349.
46 For more on LaRue and the significance of Wolverine, see Jespersen, "Unmapping Adventure." For a discussion of Angel and Wolverine in *Solar Storms*, see Castor, "Claiming Place in Wor(l)ds."

Works Cited

Ach, Jada. *Sand, Water, Salt: Managing the Elements in Literature of the American West, 1880 1925*. Texas Tech University Press, 2021.

Alaimo, Stacy. *Bodily Natures: Science, Environment, and the Material Self*. Indiana University Press, 2010.

Amimoto Ingersoll, Karin. *Waves of Knowing: A Seascape Epistemology*. Duke University Press, 2016.

Arnold, Ellen L. "Maps Over the Face of God: Remapping Epistemologies in Linda Hogan's *Solar Storms*." *Native American Literature: Boundaries and Sovereignties* 15 (2001): 49–60.

Barker, Joanne. *Red Scare: The State's Indigenous Terrorist*. University of California Press, 2021.

Bellfy, Phil. *Honor the Earth: Indigenous Responses to Environmental Degradation in the Great Lakes*. Ziibi Press, 2014.
Bilharz, Joy A. *The Allegany Senecas and Kinzua Dam: Forced Relocation through Two Generations*. University of Nebraska Press, 2002.
Bruyneel, Kevin. *Settler Memory: The Disavowal of Indigeneity and the Politics of Race in the United States*. University of North Carolina Press, 2021.
Byrd, Jodi A. *The Transit of Empire: Indigenous Critiques of Colonialism*. University of Minnesota Press, 2011.
Cardinal, Colleen. *Ohpikiihaakan-ohpihmeh (Raised somewhere else): A 60s Scoop Adoptee's Story of Coming Home*. Roseway Publishing, 2018.
Castor, Laura Virginia. "Claiming Place in Wor(l)ds: Linda Hogan's *Solar Storms*." *MELUS: Multi-Ethnic Literature of the U.S.* 31, no. 2 (Summer 2006): 157–180.
Diaz, Vicente M., and J. Kehaulani Kauanui. "Native Pacific Cultural Studies on the Edge." *The Contemporary Pacific* 13, no. 2 (2001): 315–342.
Estes, Nick. *Our History Is the Future: Standing Rock Versus the Dakota Access Pipeline, and the Long Tradition of Indigenous Resistance*. Verso Books, 2019.
Estes, Nick, and Jaskiran Dhillon. *Standing with Standing Rock: Voices from the #NoDAPL Movement*. University of Minnesota Press, 2019.
Grewe-Volpp, Christa. "The Ecological Indian vs. the Spiritually Corrupt White Man: The Function of Ethnocentric Notions in Linda Hogan's *Solar Storms*." *Amerikastudien/American Studies: A Quarterly* 47, no. 2 (2002): 269–283.
Harris, Melanie L. *Ecowomanism: African American Women and Earth-Honoring Faiths*. Orbis, 2017.
Hau'ofa, Epeli. *We Are the Ocean: Selected Works*. University of Hawai'i Press, 2008.
Hogan, Linda. *Solar Storms*. Simon & Schuster, 1995
Hornig, James F. *Social and Environmental Impacts of the James Bay Hydroelectric Project*. McGill-Queen's Press, 1999.
Jespersen, T. Christine. "Unmapping Adventure: Sewing Resistance in Linda Hogan's *Solar Storms*." *Western American Literature* 45, no. 3 (2010): 275–300.
Krech III, Shepard. *The Ecological Indian: Myth and History*. W. W Norton & Co., 1999.
Lass, William E. *Minnesota's Boundary with Canada: Its Evolution Since 1783*. Minnesota Historical Society Press, 1980.
Lawson, Michael. *Dammed Indians: The Pick-Sloan Plan and the Missouri River Sioux, 1944–1980*. University of Oklahoma Press, 1994.
Lawson, Michael. *Dammed Indians Revisited: The Continuing History of the Pick-Sloan Plan and the Missouri River Sioux*. South Dakota Historical Society Press, 2009.
Luby, Brittany. *Dammed: The Politics of Loss and Survival in Anishinaabe Territory*. University of Manitoba Press, 2020.
Maxwell, Krista. "Ojibwe Activism, Harm Reduction and Healing in 1970's Kenora, Ontario: A Micro-History of Canadian Settler Colonialism and Urban Indigenous Resistance." *Munk School of Global Affairs Working Papers Series 2009–10* (2009): 1–14.
McCutcheon, Sean. *Electric Rivers: The Story of the James Bay Project*. Black Rose Books, 1991.
Mies, Maria, and Vandana Shiva. *Ecofeminism*. Zed Books, 1993.
Moreton-Robinson, Aileen. *The White Possessive: Property, Power, and Indigenous Sovereignty*. University of Minnesota Press, 2015.
Morton, Timothy. *Ecology without Nature*. Harvard University Press, 2007.
Plumwood, Val. *Feminism and the Mastery of Nature*. Routledge, 1993.
Richard, Holly Jean, and Paul Formisano. " 'It Was the River': Indigenous Anti-Dam Literature of the Great American Desert." In *Reading Aridity in Western American Literature*, edited by Jada Ach and Gary Reger, 141–168. Lexington Books, 2020.

Royer, Marie-Jeanne S. *Climate, Environment and Cree Observations: James Bay Territory, Canada*. Springer, 2016.
Schultermandl, Silvia. "Fighting for the Mother/Land: An Ecofeminist Reading of Linda Hogan's Solar Storms." *Studies in American Indian Literatures* 17, no. 3 (2005): 67–84.
Spice, Anne. "Fighting Invasive Infrastructures: Indigenous Relations Against Pipelines." In *Paths to Prison: On the Architecture of Carcerality*, edited by Isabelle Kirkham-Lewitt, 307–358. Columbia University Press, 2020.
Stacks, Geoffrey. "A Defiant Cartography: Linda Hogan's *Solar Storms*." *Mosaic: An Interdisciplinary Critical Journal* 43, no. 1 (2010): 161–176.
Stevenson, Allyson. *Intimate Integration: A History of the Sixties Scoop and the Colonization of Indigenous Kinship*. University of Toronto Press, 2020.
Todd, Zoe. "An Indigenous Feminist's Take on the Ontological Turn: 'Ontology' Is Just Another Word for Colonialism." *Sociology Lens* 29 (2016): 4–22. https://doi.org/10.1111/johs.12124.
Todd, Zoe. "Refracting the State through Human-Fish Relations: Fishing, Indigenous Legal Orders and Colonialism in North/Western Canada." *Decolonization: Indigeneity, Education & Society* 7, no. 1 (2018): 60–75.
Toso, Tricia. " 'Keeping the Road Clear between Us': Indigenous Infrastructure and the Potential for Transformative Design." *Stream* 10, no. 1 (2018): 12–25.
Tuck, Eve, and K. Wayne Yang. "Decolonization Is Not a Metaphor." *Decolonization: Indigeneity, Education & Society* 1, no. 1 (2012): 1–40.
Warren, Karen J. *Ecofeminist Philosophy: A Western Perspective on What It Is and Why It Matters*. Rowman & Littlefield, 2000.
Whyte, Kyle Powys, Michael Dockry, William Baule, and Dean Fellman. "Supporting Tribal Climate Change Adaptation Planning through Community Participatory Strategic Foresight Scenario Development." In *Project Reports*, edited by D. Brown, W. Baule, L. Brilie, and E. Gibbons, 2014. https://glisa.umich.edu/wp-content/uploads/2021/02/GLISA_ProjRep_Strategic-Foresight.pdf.
Whyte, Kyle Powys, Nicholas J. Reo, Deborah McGregor, M. A. (Peggy) Smith, James F. Jenkins, and Kathleen A. Rubio. "Seven Indigenous Principles for Successful Cooperation in Great Lakes Conservation Initiatives." In *Biodiversity, Conservation and Environmental Management in the Great Lakes Basin*, edited by Eric Freedman and Mark Neuzil, 182–194. Routledge, 2017.
Willow, Anna J. *Strong Hearts, Native Lands: The Cultural and Political Landscape of Anishinaabe Anti-Clearcutting Activism*. State University of New York Press, 2012.

11 Deuter-Agonies

The Unmanageable Life of Alice Sakaguchi in Hiroshi Nakamura's *Treadmill*[1]

Mika Kennedy

Midway through Hiroshi Nakamura's novel, *Treadmill* (1996), 16-year-old Alice Sakaguchi is raped at a school dance. At the time, she is incarcerated at Poston, one of ten camps the United States built to confine over 120,000 people of Japanese descent during World War II.[2] Soon after her rape, Alice goes missing in the surrounding Arizona desert. By the time she is found, she is only a body. Alice's death by exposure upsets the triumphal narrative of Japanese American perseverance in the face of incarceration. It also upsets the settler-colonial narrative that incarcerated Japanese Americans tamed the desert frontier—and, through this process, reified both their own American loyalty and US claims to the desert West. Alice is unable to manage the demands of settler incorporation, which suggests tragedy; at the same time, however, her narrative also cannot *be* managed by the carceral settler structures that underpin Poston.

Nor can her narrative be managed by *Treadmill* itself, because *Treadmill* is not about Alice Sakaguchi. After her death, the novel quickly returns to protagonist Teru Noguchi's ongoing dealings with the now-infamous Loyalty Questionnaire, a more conventional narrative mechanism by which to narrate Japanese American incarceration. Meanwhile, *Treadmill* relegates Alice to "the four or more unfortunate 'incidents'" that occur in places like Poston.[3] *Treadmill*, however, is also no stranger to exhumation: although Nakamura wrote his manuscript while being incarcerated during WWII, *Treadmill* was not published until 1996, after anthropologist Peter Suzuki rediscovered the manuscript in the National Archives. *Treadmill* most recently appeared in *The Literature of Japanese American Incarceration* (2024).[4] This anthology specifically excerpts Alice's disappearance and death, which suggests that what happens to Alice is more than incidental to *Treadmill*—and to the literature of Japanese American incarceration.

Taking this invitation, I read Alice Sakaguchi in dialogue with the groundswell of recent scholarship that interrogates the settler-colonial valences of the United States' Japanese American incarceration project.[5] I argue that the Loyalty Questionnaire, which sought to separate loyal Americans of Japanese ancestry from disloyal "enemy aliens," functioned as a tool of both body and land management, and that, more than an acute response to "wartime

DOI: 10.4324/9781003581185-14

hysteria," the act of proving US loyalty also insinuates long-term incorporation into the US settler state via land labor and resettlement. Through their relationships with the Loyalty Questionnaire, both protagonist Teru Noguchi and her ghostly deuteragonist, Alice, serve key functions in connecting narratives of Japanese American incarceration to settler-colonial land management: Teru's decisions emphasize the impossibility of true agency under conditions of incarceration, and Alice Sakaguchi's "unmanageable life" embodies an alternative to settler incorporation. Against attempts to claim agency and manage land, an "unmanageable life" embraces the failure to thrive. Ultimately, Alice's "unmanageable life" presents an alternative relation to land and rethinks Japanese American narratives of belonging under conditions of incarceration, where belonging is too often assumed synonymous with incorporation into the US settler state.

Land, Loyalty, and the National Larder

To understand Japanese American wartime incarceration as a project of land and land labor is ultimately to understand the grasping extent of its carcerality—that is, the ways the carceral is entangled with the settler colonial. The War Relocation Authority (WRA) was so invested in the notion of land transformation that even though the bombing of Pearl Harbor served as alleged flashpoint for the creation of the Western Exclusion Zone and Japanese Americans' forced removal from it, WRA administrators publicly denied the camps were carceral at all. While many—both within the US federal government and in the public streets—decried Japanese Americans as unassimilable "enemy alien" threats,[6] WRA Director Dillon S. Myer insisted that most second-generation Japanese Americans—who made up the majority of those incarcerated—had "absorbed Americanism almost as naturally as they breathe."[7] Implausibly, Myer further asserted that Japanese Americans in the camps "were not prisoners—they were not internees"[8] and that the "[r]elocation centers were never intended as concentration camps or prisons."[9] But whether Myer's liberal politics understood them as prisons or not, the camps nevertheless presented a captive labor force, and Myer saw an opportunity to exploit Japanese American prisoners for nation-building purposes. Such was the vision of the camps: while imprisoned, Japanese Americans would engage in public works for public benefit, such as "land subjugation, flood control, and development of natural resources."[10] At Poston, specifically, Japanese Americans were projected to develop "80,000 acres of land for irrigation."[11] In this vision, which emphasized the ways the camps might make "maximum use of evacuee manpower,"[12] wartime manifested less as a question of military necessity than a reason to prioritize production efficiency. In addition to supporting the war effort by producing their own food, incarcerated Japanese American labor would also turn "the production of this land into the national larder."[13] Thus, as the US federal government conjured incarceration camps from the desert, the camps also manifested a broader national narrative of land made productive, frontiers reiteratively claimed.

The camps' ability to contribute to the national larder extended beyond land. WRA administrators claimed that its policies "encourage[d] the evacuee themselves to develop the highest degree of community service."[14] By adopting a classic US nation-building script, where "Americanness" is achieved through taming the frontier, Japanese Americans could cultivate their belonging within and loyalty to the United States. Thy Phu has theorized the additional valences this cultivation has for Japanese American incarceration: cultivation shares roots with notions of culture, and agricultural cultivation becomes a means of cultivating citizenship.[15] Ergo, the most productive labor—the "maximum use of evacuee manpower"—entailed work that could be seen as productive not only in the material sense of land subjugation but also in the societal sense of contributing to patriotic development. In addition to producing "American" land, incarcerated Japanese Americans became the products of American land. While this purported opportunity seems to contrast with long-standing policies in the United States of Asian exclusion on the basis of unassimilability, the reawakened frontier logics at play in the WRA's vision do more to advance the United States's settler-colonial claim to land than offer a pathway toward Japanese American belonging.

Treadmill's title functions as an address to the paradox and anticipates the political arc of the novel as a whole, given that the treadmill first emerged as a prison technology in Britain in the late eighteenth century.[16] Moreover, David H. Shayt argues that although treadmills featured in some US prisons, they were never as popular as in Britain, because the United States saw the treadmill as "fundamentally labor-*wasting*" (emphasis original).[17] The novel's title therefore serves as potent anti-metaphor—against the WRA's denial of the camps' carcerality and against the myth of the camps' productivity. To this second point, *Treadmill*'s third-person narration occasionally wanders into an unattributed, collective first-person that captures the cynicism of the Japanese Americans at Poston when offered the WRA's rhetorical frontiers:

> Just about the time we get things fixed up, we'll be shoved out. Do you think that million dollar canal they're building from the Colorado River is for us? Not a chance! By the time all the improvements bear fruit, the barracks will have fallen to pieces.[18]

Poston is only a waypoint in an ongoing string of forced removals. The land improvements are not for their benefit, because the land is not theirs. The WRA stated from the beginning that they would heed the Western states' insistences that "the evacuees be prevented from acquiring land, that the evacuees be withdrawn when the war is over."[19] Rather than return political agency to Japanese Americans through access to the frontier, land labor commodified Japanese American political life in the service of a United States that did not serve Japanese Americans, a process of alienation that Joshua T. Anderson critiques in this volume's entry on "management."[20] As on a treadmill, their labor fed an endless cycle, rather than reap just deserts.

But what is a just desert, for an immigrant community laboring within a settler-colonial state? At first blush, the injustice of the treadmill that is Japanese American incarceration seems to be Japanese Americans' inability to access the settler fantasy. Yet, the fantasy itself is an injustice. While Japanese American labor was intended for public benefit—as opposed to the private benefit of Japanese Americans—the US government intentionally sited the camps in the interior West, which was not only outside the Western Exclusion Zone but also squarely at the frontiers of the United States' ongoing project of Indigenous forced removal and genocide. The reminder is particularly sharp at Poston, which was built on the federally recognized lands of the Colorado River Indian Tribes and against the desires of the tribal council.[21] Kristen Brown further examines the "settler-colonial civilization discourse," underpinning this type of federal overreach in her writing in this volume.[22] While the WRA championed Poston irrigation work as benefitting Indigenous peoples, this does not justify the force with which the United States implemented the policy, nor its cost to Indigenous sovereignty. The incorporation of Japanese American labor into broader US settler projects typifies what Quynh Nhu Le has termed the "settler racial hegemony" of immigrant labor in the service of the settler state.[23] Attention to the settler-racial hegemonies at play recasts the WRA's counterintuitively progressive view of Japanese Americans as non-prisoners and the camps as non-prisons. In this way, land and labor become part and parcel in the story of Japanese American incarceration, and incarcerated Japanese Americans become imbricated with the US settler-colonial project. While the WRA's emphasis on land subjugation and agricultural production does not make Poston any less a prison, it awakens the settler-colonial dimensions of Japanese American incarceration. The carceral and settler colonials index their mutual entrenchment in service to one another. They also charge readers of this history and its fictions to carefully consider the ways resistance against white supremacy and xenophobia may itself be vulnerable to settler incorporation.

"The Best Ambassador I Know"

In *Treadmill,* the mechanism by which protagonist Teru Noguchi is hailed into settler incorporation—and by which Alice Sakaguchi is refused the option—is one and the same: the Loyalty Questionnaire. In 1932, versions of the Questionnaire were sent to incarcerated Japanese American adults.[24] The Questionnaire sought to make the amorphous notion of loyalty materially legible by defining it as willingness to serve in the US military (Question 27) and willingness to forswear loyalty to the Emperor of Japan (Question 28).[25] The Questionnaire and the loyalty investigation process serves as the primary driver of *Treadmill*'s plot, as Teru and her fellow Japanese Americans consider these questions' philosophical implications as well as the questions' practical exigencies, because managing land has always included controlling whose bodies are able to cross, claim, and know it.

Although originally geared toward recruitment for US military service, a far greater proportion of Japanese Americans experienced the Questionnaire as a means of expediting broad-scale resettlement and further "Americanization" outside of the camps.[26] Demonstrating loyalty by answering "yes-yes" to the questions translated to resettlement outside barbed wire, in the eastward United States. Disloyalty resulted in further sequestration at the Tule Lake segregation camp—and, for some, deportation to Japan. Ergo, the Loyalty Questionnaire, was a management tool with geographic consequences, not just in defining loyalty in terms of national allegiance but also in managing the ultimate locations of Japanese American bodies.

Working within the binary structure of the Loyalty Questionnaire, Teru Noguchi seems handpicked to embody the patriotic Japanese American that the WRA sought to identify and further cultivate. In the novel's opening scene, Teru steps off her junior college bus and into (what will soon no longer be) her family's fields in California. Her Japanese lineage manifests in her "inevitably brown eyes," but her legs are long "for a Japanese."[27] Her body is simultaneously distilled to the essentialist characteristics of her race—an essentialism more innocuous than "enemy alien" but no less potent—while also suggestive of her potential to exceed these limitations. Her legs are already longer than her racial script. Teru's genetics may be inevitable, but their meaning is mutable, and in this mutability lies a whisper of ascendancy. She could be *more*. She is pretty and educated, and if she has inevitably Japanese blood, she is still American-made. Indeed, even after her forced removal and incarceration at Poston, Teru affirms that she is "willing to serve as a WAAC if *Niseis* were allowed to join, and she was unreservedly loyal to the United States."[28] Soon, a camp administrator offers Teru a coveted secretarial job in the Midwest, earmarked for someone who might provide a deft hand and a docile voice as part of Japanese American resettlement. The administrator assures Teru, "[Y]ou'll make the best ambassador I know."[29]

Despite her ambassadorial potential, however, Teru responds *no-no* to the Loyalty Questionnaire. Rather than act as cultivated ambassador or national asset, Teru finds national loyalty irrelevant in the face of more practical exigencies: she simply does not want to be separated from her parents.[30] The practical geographic considerations attached to the Questionnaire—whose bodies will be where—outweigh any nationalist sense of how that "where" is defined. Throughout the process, Teru claims that her patriotism is undiminished; the problem is, she is not free to act on it. While the Loyalty Questionnaire ostensibly opens a door into acceptance and freedom, Teru's case functions as a reminder that there can be no true agency under conditions of carcerality.

To some degree, the ways Teru's reasoning cannot be made legible to the logics of the Loyalty Questionnaire clarify the ways Teru's life—Japanese American life—exceeds the imaginative possibilities of patriotic assimilation and settler fantasy. Her reasoning also minimizes the explicitly nation-building aspect of the WRA's land cultivation project. And yet, even as she shifts from all-American ambassador to *no-no* renunciant, Teru also remains a product of/productive of the US settler state. The Japanese American Evacuation and Resettlement Study (1946–1954), led by white sociologist Dorothy Swaine Thomas, labeled Japanese Americans who responded with *no-no* to the Loyalty Questionnaire as "spoilage," which implies their inability to successfully incorporate into the US state.[31] Teru's excision concretizes the settler-colonial state with or without her; she is excised, but her absence still reifies that which *can* purportedly belong—and by extension, what does not. Even extracted as spoilage, Teru's narrative is manageable or recapturable by the narrative of the settler state.[32]

Part of what makes Teru's narrative vulnerable to recapture is the limitations of her own critique. Teru's resistance to the logics of the Loyalty Questionnaire does not extend to the settler state. While Teru recognizes Japanese Americans' vulnerability to exploitation, this recognition does not extend to the Indigenous peoples who share this vulnerability and whose reservation lands were commandeered to build the camp; she is unable to consider the ways her story might be entangled with the settler colonial as well as the carceral. As Teru boards the train that will take her from Poston to the Tule Lake segregation camp to prepare for deportation, she takes in the landscape: "It didn't seem possible that Indians could still be living in the dilapidated shacks she saw over the fields of stunted maize. For some reason, she'd always pictured Indian reservations as well kept because they were under government protection."[33] She believes that the desert surrounding Poston is "stunted," contrary to notions of a cultivated American frontier; she does not, however, see the desert's vitality beyond these frameworks or its potential as Indigenous relative. The shock of the poor conditions of the Colorado River Indian Reservation underscores the conditions of her own experience of "government protection," but although Teru knows that the forced removals of Indigenous peoples resulted in mass death, she still chafes at the fact that all she's learning is "about Indians and the surrounding country." What point is there in learning this? One of her peers warns, "[m]aybe they're trying to make Indians out of us," heavily implying that becoming "Indian" is best avoided.[34] When defining Japanese American fortune against those of neighboring Indigenous peoples, even leaving camp is not the end of an incarceration, but the continuation of a settler-colonial management project.

The Deuter Agonies Next Door

If Teru's protagonistic subjecthood—and her agencies therein, however caveated—is initially captured by her willingness to serve, Alice serves as *Treadmill*'s deuteragonist by virtue of things to which she is unwillingly subjected. Rather than act as agonist, Alice is subjected to agony. At 16, Alice is ineligible to respond to the Loyalty Questionnaire. And while Teru is introduced by her promisingly "long legs for a Japanese," Alice's "snaky hips" spell ruin.[35]

At the school dance, juxtaposed against fellow *nisei* who ask each other, "Will YOU volunteer?" as they learn that "[f]ully half of the early registrants answered in the negative to question 28,"[36] Alice pleads *no* as an older boy, Pete Kurasaki, closes the door behind him. "I want to go back," Alice sobs. But Pete reasons, "You don't want to go back all out of breath." He pulls her down and cups her breasts and he warns her again, "People will hear you."[37]

No one does.

In the aftermath, Alice's agony is minimized amidst the political concerns of the Loyalty Questionnaire. The choice she hadn't made with Pete in that "darkened barrack" is overwritten by another choice she is not allowed to make.[38] She is too young to choose her loyalty but old enough to be incarcerated. Then, whether she has chosen this or not, Alice is pregnant:

> Three weeks later, Alice wrote a letter to the address Jim Yano gave her. She told Pete she was pregnant and was afraid to tell anyone.
>
> Enemy alien, 4-c, classifications began to pour into camp for *Nisei.* Ito, the JACL bigshot, indignantly protested to the War Department against the new classification.[39]

In the span of a line break, Alice disappears. She is displaced by rumors that "registrants would be asked to swear allegiance to the United States as a measure of loyalty."[40] The Questionnaire moves her rapist, too: he has already found work on a cattle ranch outside the camp and refuses to return. "If I stay out, I don't have to get mixed up with that cockeyed questionnaire," he claims tersely."[41] Again and again, Alice's rape is invisibilized in light of the "bigger" problem of the Loyalty Questionnaire. Nonetheless, the line break also renders Alice hypervisible. Even as Alice is overwritten by the Questionnaire and consistently upstaged, her repeated juxtaposition against the Loyalty Questionnaire entangles the two narratives more than it separates them. The sudden disjuncture of the line break keens: *what has just been lost?*

The trouble only deepens, however, when one attempts to answer. If Teru had accepted her secretarial ambassadorship and *Treadmill* had celebrated her, Alice might come across as a cautionary tale. After all, Alice isn't raped by a guard or an administrator, but a fellow Japanese American. Even if conditions of carcerality were to have influenced her rapist's actions, this is not how Alice's family understands the situation. Her parents focus on bloodline over questions of consent or context; they trace the trouble to Japan. Alice's father explains that Pete Kurasaki is *eta*, those condemned for engaging in "dirty" labor, such as handling the dead—a member of Japan's untouchable caste.[42] Terror lives at home, and it might seem that Alice is doomed by that which can be traced back to Japan, whereas Teru is offered success and belonging through Americanization and settler incorporation. Such would be the tale, if the binary logic of the Loyalty Questionnaire held true. It doesn't. *Treadmill* leaves no room for the treadmill of false incorporation.

Yet, attempt to read Alice's experience as an embodied critique of the WRA's fantasy of cultivation—violence masquerading as fertility—and this falls flat, too. Does her rape index the harm latent in the Loyalty Questionnaire, the abrogation of consent implied by forced removal and incarceration? Such a reading renders Alice important only insofar as she serves a narrative function in the novel's interrogation of the Questionnaire. She stops embodying herself, in favor of serving as metaphor.[43] erin Khuê Ninh reminds us that rape should not be understood as merely epiphenomenal—that is, as only ever serving as a canary for oppressions on allegedly grander political scales.[44] Rape, incarceration, and the Loyalty Questionnaire do not exist in parallel, one highlighted only to point out the long shadow of the other; rape is not an intermediary stop on the way to surfacing the violences of the carceral state, not least because it was a reality experienced by incarcerated Japanese Americans, and one materially co-constructed alongside and as part of their incarceration.[45]

Furthermore, even if Alice were only intended as violent collateral, the narrative is cagey about whose collateral she is. Although the *eta* caste was legally abolished in 1871, Alice's father unconsciously updates the stigma to contemporary political circumstances while explaining the family's conundrum. He conflates the *eta* caste with colonized Koreans, who were separately but similarly stigmatized under Japanese imperial occupation.[46] In the shadow of war in the Pacific, the stakes of such a slippage are global. It outsources both racialized stigmatization and settler-colonial violence to imperial Japan and decenters related critiques of the United States and the WRA.

Maybe the trouble is moot. Despite how carefully Alice story is juxtaposed against the coming of the Questionnaire, it comes to an abrupt end midway through the novel. Once they find her body, Alice is never mentioned again. What follows her death is not a period of mourning but an abruption:

> They found her bruised and blood soaked body on the second day near the river opposite Camp I . . .
>
> George wrote from Gila that he hadn't volunteered but that he had signed yes-yes and would take his chances on the draft. He was surprised that Bill and Jiro had both answered no-no.[47]

One of the last things Alice asks her family before she goes missing is, "If I go away, won't it be all right?" *Treadmill* appears to offer tacit approval of her sentiment.[48] In the span of yet another line break, mass registration plans are underway, and the Loyalty Questionnaire is once again the only name on everyone's minds. Alice is hypervisible no longer. Alice falls outside the logic of the Loyalty Questionnaire, instead defined by a litany of deuter-agonies: a loyalty unaccounted for, a crime unpunished, an unwanted pregnancy, an inherited stigma. Alice's situation must be managed because if it is not, she upsets the camp narrative of cultivation; but no assertion of Americanness can spare her what has already happened to her, and there does not appear to be any other saving grace. As though the novel itself can no longer manage her, she simply ceases to exist. This is Alice's unmanageable life.

How We Handle the Dead

Hiroshi Nakamura "thought he was writing the Great *Nisei* Novel," according to Frank Abe, an editor of *The Literature of Japanese American Incarceration* (2024).[49] The novel has the markings of such a text: namely, the rights-driven political drama of the Loyalty Questionnaire and the ways Teru makes for both archetypal ambassador on the frontier and (unwitting) resister, steadfastly asserting the reality of *nisei*, second-generation Japanese Americans. Yet, what can be done about Alice, whose story slips through the cracks, but to leave her?

Ultimately, how we read Alice is a question of how we handle the dead. Alice's "unmanageable life" names both the embodied processes that lead to death and the unruliness of exceeding the narratives that seek to define one's life. Whether through the triumph of incorporation, or of radical resistance, there is still some onus on the hero of the Great *Nisei* novel to triumph, or at least to persevere—otherwise, oppression wins. Such narratives of "compulsory survivorship" also commonly attend experiences of rape, where rape "victims" are rushed to re-identify as rape "survivors" in pursuit of reclaimed agency and healing. In that context, critics have troubled the mandate to depart so quickly from the experience of lost agency and the idea that reclaimed agency is the only viable path forward.[50] In *Treadmill*, Alice's death, as a sequela of her rape, explicitly forecloses the notion of reclaimed agency. Alice does not survive. And so, although Alice "completely disappears from the novel," Abe allows, it's her story that finds its afterlife in the anthology. He says, "We wanted to give her her due."[51] I contend that reading Alice is a means of not only attending meaningfully to the disappeared but also of charting Japanese American incarceration narratives that embrace the failure to thrive, and reject the imperative to *control* a landscape, which so often functions as the lodestar of belonging.

In the desert, a corpse won't last, but decay need not be read as failure. Heidi Amin-Hong's "decompositional reading practice" offers a vocabulary by which to understand unmanageable life, death, and resistance beyond the possibility of agency. For Amin-Hong, decay is a generative process—one that ultimately "enables an Asian American environmental relationality that is more deeply attuned to the land as an archive of decay and communal regeneration."[52] Alice might seem a counterintuitive subject in this regard, because she is *not* attuned to the land. For Alice, the desert is an empty archive. She doesn't know how far she walks through the desert. "She lets the tears run unchecked" and barely registers the desert fauna—the rattlesnake she nearly steps on, the coyote she can't quite see through the mesquite.[53] She finds no answers. She sees no way of managing the deep fractures the incarceration has wrought within in her family and can't imagine enduring the scrutiny of her insular community should she bear her *eta* rapist's child. As day turns to night and Alice realizes she cannot find her way home, she looks to the desert sky, but "she couldn't even find the Big Dipper."[54] She is unable to manage the life rapidly deteriorating around her and unable to manage the desert itself. She does not seem attuned to anything. When Alice recalls tales of those lost in the desert before her, the stories boil down to chastisements over a lack of "navigation training," which implies that but for a dearth of individual skill, Alice might have lived. After all, Alice's search party is able to successfully "beat their way through the dark mesquite."[55] They tackle the desert as an adversary and come out on top. Yet, their prowess doesn't save her, nor does it free them. As bears out in Teru's interactions with the state, no amount of skill nor potential can promise liberation. This reveals the paucity of any type of "navigation training" to address a greater problem, which is that both survival and ability are so often narrated in the violent idiom of the frontier. Agency is co-opted into the ability to conquer. What, then, does Alice really fail in?

Decompositional reading asserts an alternative relationality between subject and surround. Rather than recover Asian Americans to the foreground, it prioritizes the subject's enmeshment with the land and more-than-human, thereby positioning Asian American subjects "against the violence of colonialism that remakes land and reifies human-nonhuman hierarchies."[56] Reading Alice offers a critical reminder, however, that positioning oneself against violence does not make one impervious to violence or pain. Alice's enmeshment is painful, and it happens with a thud: "She finally tripped over a protruding root to sprawl full length on the ground," and lying flat out in the desert, she stays down, "her face resting against the cooling sand."[57] As she sobs into the sand, she personifies despair and resignation, but her tears mix with earth and her body hugs the desert into which she soon will decompose. There is no surviving, no recovery for Alice. Communal regeneration does not promise individual recuperation. But lying down (and ultimately dying) also renders her unmanageable within the script of compulsory survivorship. Rather than recuperate "Asian American subjects as visible and belonging to American environments shaped by Asian labor,"[58] Alice begins to meld into the desert she cannot tame. There, while she finds no easy oneness with the more-than-human in the desert, she is able to recognize the lack. This recognition disrupts her state of "colonial unknowing," as Lisa Fink and Lydia Cagluaq Angus describe in this volume's entry on "Relationality." What Alice is missing is not the ability to conquer, but the opportunity to relate.

Moreover, this opportunity is not the personal failing of one 16-year-old, but a pervasive quality of life in the carceral settler state. Burow, Brock, and Dove write that "settler colonialism operates through ontological foreclosures that obfuscate Indigenous ontologies of land," championing false mastery in relationality's stead.[59] This bears out in Teru's fears of Indianization and the ways Teru fails to connect the physical proximity of camp and reservation to the ideological enmeshment of the carceral and settler colonial.

While Alice, too, has none of these vocabularies, the fact remains that neither do these systems have vocabulary for her—she whose position is uncapturable by the Loyalty Questionnaire, whose fate exceeds the text itself. Rather than be seized by the need for seizure, Alice lies down, cold and quiet under the mesquite, an act that follows activist Kazu Haga's notion of "the liberation of surrender" or the act of surrendering the imperative to control.[60] Alice's "unmanageable life" unsettles the WRA's prevailing narratives of land cultivation in favor of the generative potential of decay.

Failure to conquer the frontier is failure only in the eyes of the settler state. It is also a whisper: perhaps this place was never a frontier in the first place (it is Indigenous land). Perhaps the objective was never to conquer it. The objective is not the performance of setter identity or the triumph of settler incorporation. Alice's death, her body subsumed by sand, her blood soaking into the thorns of the mesquite, asks the reader to see and know this desert differently. The WRA's land fantasy relies on the maximally resilient, productive Japanese American body, as do many Japanese American modes of resistance, which seek to claim and conquer with these same productive bodies. When survival and even "spoilage" can be co-opted to serve the carceral settler state, an unmanageable life asserts that one needn't thrive—whether defined by the state or oneself—in order to unsettle the script. The land upon which Poston was built, and Japanese American labor was meant to achieve nation-building ends, is Mohave and Chemehuevi land. It was never settler frontier, Japanese American or otherwise.

Within this reality, both aspects of Japanese American "unmanageable life" are equally critical—the resistant unruliness of being unmanageable by dominant narratives *and* the state of being unable to manage one's own life; of not being in control; of being un-resilient. Alice's vulnerability cannot be skipped past and elided. It matters that Alice does not survive, that she gives up, that her experiences were such she felt she could do nothing else. It is also worth noting that failure to thrive is increasingly part of the contemporary narrative of Japanese American incarceration and that Alice is kindred with many Japanese Americans who, in the decades after World War II, have begun to expand the incarceration narrative by recognizing incarceration as a mass disabling event, which Adria Imada has outlined in her examination of "carceral disability."[61] Imada argues that these narratives "disrupted prevailing ableist representations of Japanese Americans who had mended from incarceration."[62] Alice's unmanageable life complements testimonies from the traumatized disabled, and all those whose full stories will never be told, because they cannot be rendered into narratives of triumph.

Absolving oneself of the need to thrive turns toward the possibility that liberation has roots in other soil. This soil requires "different liberatory goals";[63] it does not assume or perpetuate the ongoingness of the settler state. "Unmanageable life" rejects the frontier imperative to triumph over, conquer, and claim and reminds us that lives that do none of these things hold value. While *Treadmill* stops short of Indigenous relation or fully formulating alternative modes of Japanese American belonging—Alice hasn't come home yet—what lingers after Alice is an invitation to recognize, and even pursue, the lack. Rather than remain a story that ends with death, reading Alice's unmanageable life opens space in the narrative where Japanese American life, too, becomes unmanageable by the carceral and settler narratives of the US state.

Notes

1 Thanks to John Ramsey, Jada Ach, and Kristen Brown for their valuable editorial feedback on this piece, as well as Frank Abe, Emma Davenport, Lauren Eriks Cline, and Ithaca College's Center for Faculty Excellence for their encouragement and support. Without them, this piece would not have been published in this century.
2 Poston was one of two camps directly managed by the Office of Indian Affairs, because they were located on federally recognized Indigenous reservation lands in Arizona. A temporary federal entity, the WRA, managed the overarching system of ten incarceration camps located across the interior West, from California to Arkansas.
3 Nakamura, *Treadmill*, 59.
4 More information about *Treadmill*'s provenance can be found in Peter Suzuki's introduction to the 1996 edition. The original edition of *Treadmill* was out of print for an extended period before its section edition in 2019. See Sarah Wald, Tara Fickle, and John Beck for further writing on *Treadmill*.
5 Jodi Byrd, Iyko Day, and Quynh Nhu Le have all published academic monographs that engage this disciplinary intersection. Hana Maruyama's 2021 doctoral dissertation, "What Remains: Japanese American World War II Incarceration in Relation to American Indian Dispossession," is also a critical contribution to this scholarship.
6 Stimson, "Letter from Secretary of War Henry Stimson to Dillon S. Myer."
7 Myer, "The Truth About Relocation."
8 Myer, "One Thousandth of the Nation."
9 Myer, "The Truth About Relocation."
10 Eisenhower, "Memorandum for Members of Congress."
11 Eisenhower, "Memorandum for Members of Congress." See also Chiang, *Nature Behind Barbed Wire*.
12 Myer, "The Truth About Relocation."
13 Arensburg, "Report on a Developing Community, Poston."
14 Eisenhower, "Memorandum for Members of Congress."
15 Phu, *Picturing Model Citizens*, 54–55. Connie Chiang expresses a similar sentiment, arguing that the camps served to "procure [Japanese American] labor in the name of assimilation and patriotism" (239).
16 Shayt, "Stairway to Redemption," 912.
17 Shayt, "Stairway to Redemption," 936. Emphasis original.
18 Nakamura, *Treadmill*, 108.
19 Eisenhower, "Memorandum for Members of Congress."
20 Anderson, "Badlands Management" XXXX.
21 Burton, et al., *Confinement and Ethnicity*, 215–219.
22 Brown, "'Back to Belonging'," pg. 171.
23 Le, "Settler Racial Hegemonies," 4, 11.
24 Kashima, "Segregation Centers and Other Camps," 163. The original "loyalty questionnaire," geared toward male Nisei to determine their eligibility for

selective service, was subsequently adopted by the WRA as part of a broader loyalty investigation.

25 These questions only fueled further questioning—does a refusal to serve translate automatically to feelings of disloyalty, particularly when one's rights of citizenship have not been recognized? Does a renunciation of loyalty imply such loyalty had ever existed?

26 Myer, "The Truth About Relocation."

27 Nakamura, *Treadmill*, 1–2.

28 Nakamura, *Treadmill*, 154. WAAC is an acronym for the Women's Army Auxiliary Corps. *Nisei* means "second-generation" in Japanese—in the U.=S context, *nisei* refers to birthright US citizens born to immigrant Japanese parents.

29 Nakamura, *Treadmill*, 161–162.

30 Teru shares this perspective with many historical *no-no* respondents. Kashima, "Segregation Centers and other Camps," 164. See also Ngai, "The World War II Internment of Japanese Americans and the Citizenship Renunciation Cases," 200. While there were many *no-no* resisters whose Loyalty Questionnaire responses were motivated by an explicit rejection of false incorporation, Teru's responses are not motivated by such clarity of political agency.

31 Thomas and Nishimoto, *The Spoilage*.

32 Elpida Ziavra's chapter in this volume, "Wastelands of Decolonial Resurgence: Managing Land and Refuse/al in Gerald Vizenor's and Leanne Betasamosake Simpson's Indigenous Narratives," examines Indigenous pushback against the designation of land and life as "waste" or spoilage.

33 Nakamura, *Treadmill*, 193.

34 Nakamura, *Treadmill*, 79, 89. Ironically, "Indianization" was a danger rather than a goal from the WRA's perspective, as well. Their eagerness to resettle Japanese Americans was partly motivated by the potential to incorporate Japanese Americans into US settlement, or colonization of Indigenous land, rather than create another "Indian problem." See Turner, "First Lady Here to Visit Hospitals."

35 Turner, "First Lady Here to Visit Hospitals," 120.

36 Turner, "First Lady Here to Visit Hospitals," 132.

37 Turner, "First Lady Here to Visit Hospitals," 122–123.

38 Turner, "First Lady Here to Visit Hospitals," 122.

39 Turner, "First Lady Here to Visit Hospitals," 125. JACL is an acronym for the Japanese American Citizens League, a Japanese American civil rights organization established in 1929.

40 Turner, "First Lady Here to Visit Hospitals," 127.

41 Turner, "First Lady Here to Visit Hospitals," 127.

42 Nakamura misspells *eta* as "*etta*" throughout the novel.

43 Pegues, "Rape Is/Not a Metaphor," 11. Pegues argues that

> experiencing sexual assault is at once emblematic and constitutive of larger structures of dominance but also remains a site of personal violence that cannot be fully apprehended by a systemic framework. . . . Rape is, therefore, always and never structural, at once and impossibly a metaphor for other violences.

44 Ninh, "Without Enhancements," 71.

45 Wallace, "Sexual Violence, Silence, and Japanese American Incarceration."

46 Nakamura, *Treadmill*, 143, 145. Eiichiro Azuma articulates the relationship between Japanese and US settler colonialism in *In Search of Our Frontier: Japanese America and Settler Colonialism in the Construction of the Trans-Pacific World* (2019).

47 Nakamura, *Treadmill*, 153.

48 Nakamura, *Treadmill*, 144.
49 Abe, personal communication via Zoom on August 8, 2024.
50 Larson, "Survivors, Liars, and Unfit Minds," 682. Much of the discourse around experiences of rape—both in terms of victimhood and survivorship, and the backlash against compulsory survivorship—comes well after Alice's rape in the 1940s.
51 Abe, personal communication.
52 Amin-Hong, "From Yosemite to the Cold War," 421–422.
53 Nakamura, *Treadmill*, 148.
54 Nakamura, *Treadmill*, 149.
55 Nakamura, *Treadmill*, 147.
56 Amin-Hong, "From Yosemite to the Cold War," 421–222.
57 Nakamura, *Treadmill*, 148.
58 Amin-Hong, "From Yosemite to the Cold War," 411.
59 Burow, Brock, and Dove, "Unsettling the Land," 58.
60 Haga, *Fierce Vulnerability,* 156.
61 Imada, "The Potential That Was in All of Us," 21, 28.
62 Imada, "The Potential That Was in All of Us," 23.
63 Arvin, Tuck, and Morrill, "Decolonizing Feminism," 16.

Bibliography

Abe, Frank, and Floyd Cheung, eds. *The Literature of Japanese American Incarceration*. Penguin, 2024.

Amin-Hong, Heidi. "From Yosemite to the Cold War: Decomposing Settler Mythologies in the Asian American Outdoors." *Journal of Asian American Studies* 26, no. 3 (2023): 405–426.

Arensberg, Conrad M. "Report on a Developing Community, Poston." California State University, Sacramento, Department of Special Collections and University Archives [online version available through the Densho Digital Repository (object ID ddr-csujad-55-1854). Accessed August 7, 2024. www.ddr.densho.org.

Arvin, Maile, Eve Tuck, and Angie Morrill. "Decolonizing Feminism: Challenging Connections Between Settler Colonialism and Heteropatriarchy." *Feminist Formations* 25, no. 1 (2013): 8–34.

Azuma, Eiichiro. *In Search of Our Frontier: Japanese America and Settler Colonialism in the Construction of Japan's Borderless Empire*. University of California Press, 2019.

Burow, Paul Berne, Samara Brock, and Michael R. Dove. "Unsettling the Land." *Environment and Society* 9 (2018): 57–74.

Burton, Jeffery F., Mary M. Farrell, Florence B. Lord, and Richard B. Lord. *Confinement and Ethnicity: An Overview of World War II Japanese American Relocation Sites*. University of Washington Press, 2002.

Beck, John. "Dust Breeding: Narratives of Inter(n)ment." In *Dirty Wars: Landscape, Power, and Waste in Western American Literature*, 73–100. University of Nebraska Press, 2009.

Byrd, Jodi. *Transit of Empire: Indigenous Critiques of Colonialism*. University of Minnesota Press, 2011.

Chiang, Connie. *Nature Behind Barbed Wire: An Environmental History of the Japanese American Incarceration*. Oxford University Press, 2018.

Day, Iyko. *Alien Capital: Asian Racialization and the Logic of Settler Colonial Capitalism*. Duke University Press, 2016.

Eisenhower, Milton S. "HST-SVP: Harry S. Truman Papers as U.S. Senator and Vice President." Harry S. Truman Library, Independence, MO [online version available

through the Archival Research Catalog (ARC identifier 602137)]. Accessed August 7, 2024. www.archives.gov.
Espiritu, Yen. *Body Counts: The Vietnam War and Militarized Refuge(es)*. University of California Press, 2014.
Fickle, Tara. "*No-No Boy*'s Dilemma: Game Theory and Japanese American Internment Literature." *Modern Fiction Studies* 60, no. 4 (2014): 740–766.
Haga, Kazu. *Fierce Vulnerability*. Parallax Press, 2025.
Imada, Adria L. "'The Potential That Was in All of Us': Carceral Disability and the Japanese American Redress Movement." *WSQ: Women's Studies Quarterly* 51, nos. 1 & 2 (2023): 21–47.
Kashima, Tetsuden. "Segregation Centers and Other Camps." In *Judgment Without Trial: Japanese American Imprisonment During World War II*, 160–192. University of Washington Press, 2004.
Larson, Stephanie R. "Survivors, Liars, and Unfit Minds: Rhetorical Impossibility and Rape Trauma Disclosure." *Hypatia* 33, no. 4 (2018): 681–699.
Le, Quynh Nhu. "Settler Racial Hegemonies: The Tense Crossings of Settlement, Empire, and Race." In *Unsettled Solidarities: Asian and Indigenous Cross-Representations in the Américas*, 1–24. Temple University Press, 2019.
Maruyama, Hana. "What Remains: Japanese American World War II Incarceration in Relation to American Indian Dispossession." PhD diss. University of Minnesota, 2021.
Myer, Dillon S. "HST-DSM: Dillon S. Myer Papers." Harry S. Truman Library, Independence, MO [online version available through the Archival Research Catalog (ARC identifier 602137)]. Accessed August 7, 2024. www.archives.gov.
Nakagawa, Martha. "False Constructions of Loyalty: The Real Resistance against Incarceration." In *John Okada: The Life and Rediscovered Work of the Author of No-No Boy*, edited by Frank Abe, Greg Robinson, and Floyd Cheung, 277–283. University of Washington Press, 2018.
Nakamura, Hiroshi. *Treadmill*. Mosaic Press, 1996.
Ngai, Mae. "The World War II Internment of Japanese Americans and the Citizenship Renunciation Cases." In *Impossible Subjects: Illegal Aliens and the Making of Modern America*, 175–201. Princeton University Press, 2004.
Ninh, erin Khuê. "Without Enhancements: Sexual Violence in the Lives of Everyday Asian American Women." In *Asian American Feminisms and Women of Color Politics*, edited by Lynn Fujiwara and Shireen Roshanravan, 69–81. University of Washington Press, 2018.
Pegues, Juliana Hu. "Rape Is/Not a Metaphor." *Journal of Asian American Studies* 24, no. 1 (2021): 9–17.
Phu, Thy. *Picturing Model Citizens: Civility in Asian American Visual Culture*. Temple University Press, 2011.
Shayt, David H. "Stairway to Redemption: America's Encounter with the British Prison Treadmill." *Technology and Culture* 30, no. 4 (1989): 908–938.
Thomas, Dorothy S., and Richard Nishimoto. *The Spoilage: Japanese American Evacuation and Resettlement*. University of California Press, 1947, 87. Accessed June 21, 2024. https://archive.org/details/spoilagejapanese00thom.
Turner, Timothy. "First Lady Here to Visit Hospitals." *Los Angeles Times*, April 27, 1943.
Wald, Sarah D. "The 'Clouded Citizenship' of Rooted Families: Japanese American Agrarianism in *Rafu Shimpo, Kashu Mainichi*, and *Treadmill*." In *The Nature of California: Race, Citizenship, and Farming Since the Dust Bowl*, 75–101. University of Washington Press, 2016.
Wallace, Nina. "Sexual Violence, Silence, and Japanese American Incarceration." Accessed April 2, 2018. https://densho.org/catalyst/sexual-violence-silence-japanese-american-incarceration/.

12 Trail Reviews

Further Commodification of Wilderness

Eytan Pol

The Waterpocket Fold, the 100-mile-long monocline that comprises the majority of Capitol Reef National Park in southcentral Utah, is an eroded land of sudden angles and wild geological distortions. Perception of the land does not equate to comprehension, and one is confronted with an absurd land scape which blurs logic. Yet, it remains difficult to avert one's eyes from the wild angularity of the imposing cliffs and deep canyons. Environmentalist scholar Paul Shepard suggested in *Man in the Landscape* that people's attraction to symmetry and lines stems from an ancient and primate instinct to spot adequate and reliable trees and secure branches on which to climb.[1] When I lived and worked in Capitol Reef as an international volunteer during the first half of 2020, I had no concept of this instinct, though I believe I experienced it during those five months nonetheless.

One of my favorite trails to hike during the afternoon trail patrols in the frontcountry was the Frying Pan, named for the bowl-shaped dip in the monocline the trail traverses, as well as for the lack of shade that is available to shelter from the burning Utah sun. The Frying Pan is a connecting trail, which branches off halfway from one trail and then leads to intersect with another one roughly three miles toward the south. Thus, the trail does not start at a traditional trailhead at a site near a road or parking lot, nor does it end at a satisfactory conclusion in the form of an overlook or at least loop around itself back to its starting point. The hiker is simply dropped from one trail onto another—out of the frying pan and into the fire. This lack of a usual climax at the trail's end, a reward at the finish line, perhaps causes one to pay greater attention to the surroundings. Without a typical destination, the journey itself is the experience. Large boulders stand impossibly balanced near the bottom of the bowl in the monocline halfway through the trail. Differently colored rock layers contrast starkly and then suddenly swap places in unexpected angles. A modest juniper guards the sandy wash that the trail crosses at one point. A raven once suddenly flew away from this juniper as I passed, startling me as I startled it.

Not all hikers share my view of the Frying Pan. "This section was my least favorite," one hiker wrote in his three-out-of-five-star review of the Frying Pan Trail on *AllTrails*. "Some of the cairns were poorly kept."[2]

DOI: 10.4324/9781003581185-15

A multitude of outdoor apps focused on exploring wilderness exist, and most offer similar functionalities. *AllTrails* is one of the most well known and widely used of these apps, boasting 65 million users and more than a billion logged miles on their website. Any user can locate nearby trails on a map and subsequently obtain detailed information, such as a description of the trail, the length of the trail, the expected elevation gain, the difficulty, the average time it takes to hike the trail, the type of trail (such as a loop, an out-and-back, or a connecting trail), the condition of the trail (rocky, unmaintained, etc.), directions, weather reports, whether the area is dog friendly or not, and even the expected mosquito activity. Additionally, the app provides reviews of each trail along with an aggregated score on a five-star scale. An interactive route for each trail is included, which the user can follow during their hike to make sure they do not accidentally stray off trail. When a user starts a hike, *AllTrails* will track real-time progress, such as the time hiked, the distance covered, the elevation gained, the average pace, as well as the remaining time and distance. After a hike is finished, the user is prompted to leave a review and give their rating of the trail. Evidently, an enormous amount of information and data is provided to the user, but what is sacrificed in return?

On their website, *AllTrails* describes how they continue to be driven by the desire to share the outdoors with as many people in as many places as possible.[3] Accordingly, the company's intent is to promote outdoor recreation, while at the same time, profit is realized through the maximization of the user-base. *AllTrails* is free to use, but a subscription allows for more functionalities such as the ability to download maps for offline use and access to 3D overviews of routes. Growth and profit appear to be accomplished through the gamification of wilderness recreation. This gamification occurs in multiple ways. The app's data, such as the real-time tracking of distance and pace provided during a hike, can be seductive, and the user is tempted to beat the average time of hiking and increase their own pace. At the end of the hike, an overlook might be reached, after which the user can compare the images provided on the app with the reality on the ground, and perhaps add their own photos. Completed trails can be saved and shared and are displayed on the app. Occasionally, *AllTrails* reminds the user with notifications of their number of hiked miles in a particular month, which urges them to increase this number.

In turn, this gamification of wilderness might lead the user to seek out quantity and efficiency over more experiential factors. After all, why hike the Frying Pan if a shorter trail with a higher score leading to an arch, for example, might be visited with less effort. Most of us who visit public lands or other wilderness areas have a limited amount of time available to us, and it therefore makes sense that we might want to use this time wisely. In other words, a tendency emerges to maximize experiences in the wilderness through efficiency and planning. As a result, the nearby Cassidy Arch Trail with its shorter length and impressive overlook attracts more visitors than the Frying Pan toward the north. Cassidy Arch has a higher rating and, at the time of writing, 30 times the number of reviews as the Frying Pan, both of which nudge visitors who don't mind more crowded trails to pick the former over the latter. While hikers indeed might gain a nice view, they sacrifice a more engaged outdoor experience by following a company's suggestions, data, and algorithms.

AllTrails was launched in 2010, but it is not the only game in town, nor is it the first. *GaiaGPS*, founded in 2008, is a comparable app that focuses more on route creation and tracking. In 2012, *The Hiking Project* was founded, which is more user-based and crowdsources its hiking guides. Here, too, gamification is rife, as users and contributors earn points and are subsequently ranked with one another. *Strava* was released in 2009 and is intended for fitness in cities, though it includes mountain biking and kayaking routes in addition to hiking trails. The prevalence of comparable outdoor apps highlights the popularity of a digital and consumerist approach to outdoor recreation. This increase of consumerism and/or gamification in relation to wilderness recreation is neither abstract nor theoretical. For example, researchers are analyzing AI-powered outdoor app optimization and its comparable outdoor apps highlights the popularity of a digital and consumerist approach to outdoor recreation. This increase of consumerism and/or gamification in relation to wilderness recreation is neither abstract effects on "hiking as a tourism product."[4] Notably, this illustrative study assumes as unchallenged facts that wilderness recreation is a product, as well as the notion that AI and outdoor apps undoubtedly enhance these experiences.

The gamification and commodification of public land and wilderness recreation do not exist in a vacuum. The NPS itself promotes similar attitudes through products such as collector coins in visitor center gift shops and, in particular, the National Park Passport. Visitors can collect different coins at different parks and stamp their passport at every visitor center and thus collect and compare visitation mementos. When I worked at Capitol Reef, "where is the passport station?" was perhaps the most common (and occasionally the only question) from visitors. Through the sale of these products, the NPS appears to promote visitation quantity. Ultimately, this is understandable, as portions of a park's funding depend on annual visitation numbers. If a national park has a large increase in visitation, there is a subsequent need and expectation for more funding. As an underfunded and understaffed agency—realities that have increased dramatically through recent budget cuts—any budgetary increases are naturally desired. Therefore, gamification and commodification are not necessarily unwelcome.[5]

In some ways, these outdoor apps are a modern iteration or evolution of other information sources or manners in which to enjoy and recreate outdoors, such as paper hiking guides or trail recommendations from park rangers, which in turn have aggregated and centralized word of mouth suggestions and individual experiences of trips and explorations. However, two aspects truly set apart outdoor apps from these other sources. First, there is the incomparably vast amount of information provided in these apps. A hiking guide, even a book, can only contain so much information, and one is able to carry only so many books in a backpack. These books and other physical guides are usually focused on specific regions, sometimes even single sections of a National Forest, for example. In contrast, *AllTrails* contains detailed routes, reviews, and data about almost half a million trails globally. Moreover, this data is continuously updated and expanded. The company's title alludes to certain Leviathan aspirations: the inclusion of all trails. Both in the scope of the data provided as well as the sheer quantity, *AllTrails* is comprehensive and all encompassing to a far larger degree than any paper-hiking guide—the regional is centralized into the global. Second, and perhaps most crucially, are the economization and commodification of wilderness which occur through the app's quantifiable, standardized, and comparable data: the user's ability to rate and appraise a wilderness area and add one's score to an aggregated average available to all others. These factors, I believe, result in a loss of spontaneity as well as the diminishment of wilderness itself.

Before I discuss how these outdoor apps diminish the values of wilderness, I will outline my own understanding of wilderness. Scholars, legislators, philosophers, and others have long debated definitions of wilderness. As of 2025, almost 112 million acres of officially designated wilderness exists in the United States, but beyond legal status, there is no full overlap between the official designation and all the wild areas in the country. The Maze, a district of Canyonlands National Park in Utah, is not federally designated as wilderness, despite its existence as one of the most remote areas in the Contiguous United States—someone wandering around the deep solitary canyons and mesas of the Maze might consider such federal designations abstract. Nor does a designation of an area as a national park necessarily signify wildness; Gateway Arch National Park in St. Louis springs to mind. Thus, various definitions of wilderness, independent of legal and federal status exist.

Environmental historian Roderick Nash traced the etymological roots of "wilderness," which refer to a characteristic of being "self-willed, willful, or uncontrollable."[6] Indeed, in my view, the degree of control over an area is a crucial aspect of wilderness. Control takes many forms, though physical and material control (such as development, roads, and structures) is the most impactful and concrete kind. Contentiously, in Nash's understanding of wilderness, "the absence of men and the presence of wild animals is assumed."[7] If the presence of people is incompatible in this anthropocentric understanding of wilderness, then wilderness is impossible to define as it is logically imperceptible. This paradox effectively likens wilderness to anything outside the observable universe as the presence of the observer in wilderness immediately causes its disappearance. While it was undeniably important for land management policy, the 1964 Wilderness Act includes a similar caveat, as it adds that wilderness is a place "where the earth and its community of life are untrammeled by man, where man himself is a visitor who does not remain."[8] In this light, peoples' presence is now compatible with wilderness, as long as their stay is temporary. But how long can one visit before wilderness is no longer wild? And, as Surabhi Balachander notes earlier in this collection, what about Indigenous people who were and are present in what we commonly understand as wilderness, not as visitors, but as multigenerational inhabitants?[9] A more layered definition is needed.

Ecological philosopher Max Oelschlaeger argues that "our definitions of 'wildness' and 'wilderness' preclude recognition of nature as a spontaneous and naturally organized system in which all parts are harmoniously interrelated; in consequence, human kind has believed itself compelled to impose order on nature."[10] Oelschlaeger does not exclude people from wilderness. Rather, he emphasizes the omission of spontaneity, which leads to an imposition of order and control as a crucial part of his understanding of wilderness. Importantly, as others have suggested, control should be understood as a matter of degrees rather than a set threshold that clearly distinguishes wilderness.

This definition of wilderness, which centralizes physical or material control, is at odds with some ecocritical understandings. Environmental historian William Cronon, for example, suggests that wilderness might exist in one's backyard.[11] Cronon argues that "the tree in the garden is in reality no less other, no less worthy of our wonder and respect, than the tree in an ancient forest that has never known an ax or saw," which in turn leads him to reject the dualism between the garden as artificial and the wilderness as natural.[12] Wonder and respect aside—as these are dependent on the individual—I believe the distinction between an environment that is perhaps managed and controlled by humans, like a backyard, and a more spontaneous, self-ordered wild area is important to maintain, if only for the purpose of environmental preservation. If one truly does not differentiate between a redwood and a cultivated apple tree, for example, there is nothing to stop a logging company from cutting down the former and planting more of the latter beyond federal and legal land protection. President Donald Trump drastically reduced Bears Ears National Monument area during his first term (which was subsequently restored by President Joe Biden) showing the fragility of such federal protections. According to Cronon's understanding of wilderness, which equates the planted backyard tree to a spontaneous one, we lose nothing when a wilderness area is reduced or eliminated. The hesitancy of recognizing a distinction between an artificial environment and a wild one is perhaps because a hierarchy is implied. This need not be the case; one can still maintain wonder and respect for one's backyard without equating it to wilderness. Viewing wilderness as a spectrum influenced by material control avoids binary categorizations.

Other ecocritics question whether wilderness as it is commonly or traditionally understood is privileged in scholarship, and some even suggest letting go of the concept altogether, as it is considered to have "arisen from the Enlightenment and imperial processes."[13] I am hesitant to let go of the concept of wilderness. Perhaps this is a result of being from a densely populated country, the Netherlands, that has no wilderness left; redefining one's backyard as a wilderness appears a luxury when one's country still has remote areas and considerable public land. There is no clear breaking or taming point, at which exact moment the degree of control turns wilderness into something subdued. One signpost does not tame a canyon, nor does a single coyote turn a city park wild. Yet, one is still able to differentiate between the Grand Canyon and a petting zoo, or between a planted backyard tree and a spontaneous redwood. Diminishment of wildness and wilderness is thus a gradual process.

If an increase in physical or material control leads to a diminishment of wildness, how do outdoor apps catalyze this process? The quantity of comparable data that the apps provide and collect, I believe, leads to a process that Jack Turner notes, "reduces everything to a unit of measurement."[14] *AllTrails* does not only compile the physical circumstances and datapoints of a trail, such as distance, elevation, and time—which, of course, already derive from accompanying units of measurement such as miles, feet, and hours—but crucially, the app also reduces the appreciation of wilderness to measurable and comparable units by providing reviews, ratings, and tags. *AllTrails* utilizes standardized tags to categorize every trail, such as "[picturesque] views," "historic site," "wildflowers," and "or no shade," which are part of the foundation for its system to measure, rate, and compare trails. The ability to rate trails is continuously updated through incoming reviews and leads to the creation of a hierarchy. This is what truly sets an app like *AllTrails* apart from an NPS leaflet or the copy of Rudi Lambrechts' *Hiking the Escalante* in my bookcase, for example. According to philosopher Carolyn Merchant, economic attitudes toward wilderness stem from Cartesian and mechanistic philosophies which assumed that "nature was manageable through rational understanding and efficient action."[15] Efficiency and rationality allow the layer of economic language to be forced upon wilderness, which enables the quantifiable comparison of the aforementioned Frying Pan with any other trail from Alaska to Argentina. When something can be measured in a standardized way, it can subsequently be compared, rated and ranked, and profited from.

Mishuana Goeman states that "without labor to tame the land, it is closely assigned the designation of 'nature' or 'wilderness.' As such, property is not just a material, but it is also constructed through social relationships."[16] This social relationship is often one of commodification, and as economist Karl Polanyi states, "production is interaction between man and nature" and both, whether in the form of labor, land, or resources, would "be dealt with as commodities" if they are left to be self-organized in a market economy.[17] However, land and labor are entirely fictitious commodities, according to Polanyi.[18] The same applies to wilderness: a redwood does not sprout as a product, but is interpreted and consumed by humans as such. The convertibility to units of measurements allows this commodification. As Turner states, "an hour's labor is worth a certain amount of money; so is three hundred board-feet of redwood."[19] The unit of measurement, whether it is time, money, or something else entirely, allows a redwood to be converted into hours of labor. In other words, the measuring and commodification of nature by outdoor apps are nothing new. Instead, it is that which is commodified wherein a new leap occurs. A redwood is material; it can be touched, measured, cut down, and processed. In contrast, outdoor apps measure and commodify a part of wilderness that, while based on material circumstances, is itself entirely abstract and immaterial, such as a hike, a view, a visit, or an experience. These experiential commodities are not finite in the way that a tree only has a certain amount of board-feet or a mine has only a certain amount of ore. A view or a hike cannot be consumed in such a limited or reducible manner.

By utilizing data, tags, and ratings, *AllTrails* further commodifies and controls wilderness concretely through the filter functionality. The inclusion of numerical ratings allows the user to filter out trails below a certain ratings threshold and reject those with a lower score. Similarly, the filter can be applied to prioritize popularity, that is, trails with large quantities of reviews. Both of these options further the hierarchical consideration of wilderness as a result of their convertibility into measurable units. Additional filters exist, such as distance, difficulty, type, and a multitude of other factors. Gamification of wilderness recreation reemerges through the option to filter out completed (previously hiked) trails in an area. This suggests an attitude similar to the National Park Passports, in which a quantity of visits is seen as important—the app allows to user to make sure the same trail is not hiked twice. A consumerist approach to wilderness recreation is thus present: an area or trail is "used up" after a first visitor or consumption, like a movie ticket or a milkshake.

Particularly significant is the ability to filter by "attractions," which represent something noteworthy or breathtaking about a respective trail: a reason to hike or visit it. Examples on *AllTrails* include "waterfall," "views," "lake," "hot springs," "wildflowers," "cave," "historic site," and numerous others. Outdoor apps prioritize the expectation of seeing an attraction on a trail, and by allowing the user to filter for these specifically, picturesque scenery in the wilderness is privileged. Turner argues that "the national parks were created for, and by, tourism, and they emphasize what interests a tourist—the picturesque and the odd. They are managed with two ends in mind: entertainment and the preservation of the resource base for entertainment."[20] To a degree, this tendency to entertain is present within the NPS itself; many, though certainly not all, trails do lead toward a site that *AllTrails* might label an attraction. However, this is not the sole purpose of the agency. The NPS does promote, and often prioritize, cultural, natural, historical, and Indigenous interpretations, as well as wilderness protection and preservation. As noted earlier by Lisa Fink and Lydia Cagluaq Agnus, the NPS is "advancing efforts to collaborate with Native nations."[21] Rather, Turner's description would be more accurate when applied to outdoor apps, as these indeed recognize the wants and wishes of its user base, of which a picturesque site seems to be the primary goal. To return to the example of Capitol Reef National Park, it is no surprise that the trails that contain what *AllTrails* filters as attractions, whether these are waterfalls, arches, natural bridges, (slot) canyons, or overlooks, receive both higher ratings and have more reviews than those lacking any such singular attractive destinations or vistas, such as the Frying Pan. The option to filter for attractions might cause people to think of wilderness in terms of beauty, rather than holistically: a place where the beautiful exists alongside the bland, the harsh, and even the horrific. If only the most aesthetic parts are deemed (economically) desirable, wilderness runs the risk of becoming an outdoor museum. With such a paradigm shift, wilderness must be pretty to remain viable.

Indeed, when attractions and the picturesque are prioritized or valued most in wilderness, places with perhaps less impressive singular vistas risk being overlooked. The aesthetic locations can be digitally separated or filtered from those deemed unremarkable, and because of all the accompanying photos, the visitor can preemptively inspect the trail like a customer examining a potential purchase. This filtering reveals to tourists and business alike which areas and trails are traveled more and are thus (economically) desirable. It does not take a fanciful imagination to foresee what may happen to those places that are less visited, appreciated, and desired; from Hetch Hetchy to Glen Canyon to Bears Ears, history has shown that economic interests and the allure of profit regularly supersede environmental or preservationist interests. By intensifying the economic consideration of wilderness through the commodification of the abstract, other utilizations of wilderness may yet again become common if there is no profit to be made through visitation or tourism. What will remain is an impressive view, an *AllTrails* attraction, or the board-feet, strip mine, or pumpjack. The standardization into measurable units and commodification of wilderness recreation enable outdoor apps to create this expectation of a certain number of picturesque sites that can be visited or, more accurately, consumed.

The increasing control over wilderness is initially as abstract as the economic attitude from which it springs. However, by increasing and concentrating visitation to picturesque sites, as the ratings and reviews make possible, physical consequences of the increase of control quickly emerge. Angel's Landing, one of the most well-known trails in Zion National Park, illustrates the physical manifestation of control. Angel's Landing is most known for the last half-mile section of the trail, which leads hikers up a narrow monolith shaped like an enormous shark's fin. There is a dramatic 1,400-feet drop toward the bottom of the canyon on both sides of the narrow trail, which is located in the heart of Zion Canyon. The view on top of Angel's Landing is undeniably captivating, the embodiment of the picturesque, and this is what attracts many visitors. The trail has a nearly perfect 4.9 out of 5 star rating on *AllTrails*, with a staggering 31,000 reviews, dwarfing both the aforementioned Cassidy Arch and Frying Pan Trails.[22]

However, besides being picturesque, Angel's Landing is also quite dangerous. Visitors occasionally fall from the narrow path and plummet to their deaths. When I first visited Angel's Landing in 2020, a warning sign noting the ten deaths that had occurred on the trail stood menacingly at the trailhead. Now five years later, the number of deaths has already almost doubled to 18. This exponential increase in deaths no doubt mirrors the exponential visitation of the trail due to the increased popularity. Images from the top of Angel's Landing attract visitors magnetically to the dangerous trail. A quick scroll through the *AllTrails* reviews shows the praises for the picturesque views, which commonly include adjectives such as magnificent, iconic, and heavenly. However, the situation regarding both the deaths and the increased visitation has become unsustainable. Cables, iron bars, and safety handholds have been drilled into the rock of the monolith years ago, and since 2022, permits are required to hike the trail. These permits are obtained through a lottery system. If a visitor is lucky enough to be randomly allocated a ticket through this lottery, they must pay a fee, after which they are allowed to hike up Angel's Landing during a reserved timeslot. To use the appropriate terminology, the demand for the trail greatly outweighs the supply. In such a case, the abstract commodity of a view is not finite in the sense that it can be used up, but only a limited number of visitors can have access to the commodity in a certain amount of time. The control of wilderness is then no longer abstract. The demand, which cannot be met, has necessitated control in the form of both physical alterations as well as entry requirements through a paid lottery system which awards permits. I do not wish to suggest that the current state of Angel's Landing can be solely blamed on outdoor apps like *AllTrails*. The apps reflect a desire that is already present, though in turn they certainly exacerbate and bolster this desire further through their system of economic appraisal quantifying wilderness. The result is measures like those present at Angel's Landing at other places.[23]

The fact that outdoor apps are not part of a federal institution or agency makes any quick solution difficult. *AllTrails* and similar companies are private companies that, in a sense, manage public land indirectly or at least influence the management by those agencies in charge of a wilderness area. Concentrating visitation on one trail necessitates a concentration of funding for that same trail. This can take the form of extra trail patrols, a prioritization of trail maintenance, or even an increase in search-and-rescue resources.[24] Additionally, these outdoor apps do not immediately impact wilderness physically, like those looting archeological sites, hunting unlawfully, driving ATVs, or mining or logging illegally do. Only through their commodification of wilderness is damage eventually done. One cannot easily write to a state representative to petition against the development of any particular wilderness area because of outdoor apps. Moreso, in the context of ecological and environmental issues, Lynn White Jr. expressed doubt that dilemmas created by technology "can be avoided simply by applying to our problems more science and more technology."[25] Thus, I believe any solution will inevitably be decentralized as no oversight exists nor are any laws broken, and neither will it be technological in nature.

I imagine that the removal of ratings and reviews from outdoor apps would have immediate positive results by eliminating the commodification through units of comparable measurements, which in turn would (partially) promote spontaneity and reverse the control over wilderness. However, it is precisely the ratings and reviews which are crucial for the functioning and growth of outdoor apps. Without these, the option to filter, the social aspect, and the gamification would all lose much of their functionality, and the apps themselves would be deprived of what truly set them apart from other sources of information of the outdoors. Moreover, as Shepard states, one must go beyond only blaming the tools themselves.[26] Again, the outdoor apps reflect a desire already present within; their commodification of wilderness does not exist wholly isolated from other economic attitudes toward wilderness.

Any place one visits has already been previously mapped, photographed, discovered, and inhabited by others, Indigenous or otherwise. As Gary Snyder writes, a wilderness without humans has not existed for thousands of years.[27] However, one can choose spontaneity, to reject and counter the increase of control over wilderness by outdoor apps. What has personally helped is to circle a place of interest on a simple roadmap, such as a forest and mountain range or National Monument, and, whenever possible, to refrain from creating schedules and careful itineraries for our time available to travel. Allowing mood and curiosity to lead us rather than alluring photographs and aggregated ratings has brought back wonder and spontaneity. Moreover, following along with a route on one's phone requires regular monitoring of one's phone, rather than pay attention to the surroundings. Not knowing what lies at the end of a trail, or what attraction may be present around the bend of a canyon or the five-star rating of that very canyon, leads to a wholly different experience of wilderness than being guided by the interests of commodification.

Notes

1 Shepard, *Man in the Landscape*, 5.
2 Oswald, review of "Frying Pan Trail," *AllTrails*.
3 *AllTrails*, "Our Values."
4 Ivanova and Wald, "Recommender Systems for Outdoor Adventure Tourism Sports," 347. This article aggregates and summarizes over a thousand research papers focused on the functioning of outdoor apps and contains detailed points of attention and functionality aspects and concerns of these apps to a far larger degree than I am able to summarize in this chapter.
5 It is important to note that an individual park ranger's view of such products might radically differ from those higher up in the NPS hierarchy who make such decisions.
6 Nash, *Wilderness and the American Mind*, 1.
7 Nash, *Wilderness and the American Mind*, 3.
8 *Wilderness Act*, 2.
9 Balachander, "Rural/Wilderness."
10 Oelschlaeger, *The Idea of Wilderness*, 8.
11 Cronon, "The Trouble with Wilderness," 89.
12 Cronon, "The Trouble with Wilderness," 88–89.
13 Fletcher, et al., "Indigenous Knowledge and the Shackles of Wilderness," 2.
14 Turner, *The Abstract Wild*, 63.
15 Merchant, *The Death of Nature*, 280.
16 Goeman, "Land as Life."
17 Polanyi, *Great Transformation*, 136.

18 Polanyi, *Great Transformation*, 76.
19 Turner, *The Abstract Wild*, 63.
20 Turner, *The Abstract Wild*, 27.
21 Fink and Agnus, "Relationality," 10.
22 For reference, Angel's Landing has 50% more reviews on *AllTrails* than the Washington Monument has on Google, a much larger platform.
23 Angel's Landing's permits and lotteries are not unique in the National Park system. Another trail in Zion National Park, the Subway, requires a similar lottery permit. Moreover, entire national parks now function on a permit system for entry for large portions of the year, such as Arches and Rocky Mountains National Parks. Yosemite has announced similar measures starting in the summer of 2025.
24 A law enforcement ranger at Capitol Reef once told me that most search-and-rescues occur not on the long, remote, and dangerous trails, but rather on two short and simple hikes to an arch and some petroglyphs, respectively. After all, this is by far where most visitors are drawn.
25 White, "The Historical Roots of Our Ecologic Crisis," 12.
26 Shepard, *Man in the Landscape*, 215.
27 Snyder, "The Etiquette of Freedom," 24.

Bibliography

Cronon, William. "The Trouble with Wilderness." In *Uncommon Ground: Rethinking the Human Place in Nature*, edited by William Cronon. Norton, 1995.

Fletcher, Michael-Shawn, Rebecca Hamilton, Wolfram Dressler, and Lisa Palmer. "Indigenous Knowledge and the Shackles of Wilderness." *Proceedings of the Natural Academy of Sciences* 118, no. 40 (2021): 1–7.

Goeman, Mishuana. "Land as Life: Unsettling the Logics of Containment." In *Native Studies Keywords*, edited by Stephanie Nohelani Teves, Andrea Smith, and Michelle H. Raheja. The University of Arizona Press, 2015.

Ivanova, Iustana, and Mike Wald. "Recommender Systems for Outdoor Adventure Tourism Sports: Hiking, Running, and Climbing." *Human-Centric Intelligence Systems* 3, no. 3 (2023): 344–365.

Merchant, Carolyn. *The Death of Nature: Women, Ecology, and the Scientific Revolution*. HarperCollins, 1989.

Nash, Roderick Frazier. *Wilderness and the American Mind*. Yale University Press, 2001.

Oelschlaeger, Max. *The Idea of Nature*. Yale University Press, 1991.

Oswald, Steven. Review of "Frying Pan Trail." *AllTrails*, February 5, 2015.

"Our Values." *AllTrails*, 2024. https://www.alltrails.com/about#our-values.

Polanyi, Karl. *Great Transformation: The Political and Economic Origins of Our Time*. Beacon Press, 2001.

Shepard, Paul. *Man in the Landscape: A Historic View of the Esthetics of Nature*. The University of Georgia Press, 2002.

Snyder, Gary. "The Etiquette of Freedom." In *The Wilderness Condition: Essays on Environment and Civilization*, edited by Max Oelschlaeger. Sierra Club Books, 1992.

Turner, Jack. *The Abstract Wild*. The University of Arizona Press, 1996.

White, Lynn, Jr. "The Historical Roots of Our Ecologic Crisis." In *The Ecocriticism Reader: Landmarks in Literary Ecology*, edited by Cheryll Glotfelty and Harold Fromm. The University of Georgia Press, 1996.

Wilderness Act, 1964, 88th Congress. https://www.fs.usda.gov/Internet/FSE_DOCUMENTS/fseprd645666.pdf.

Part 3

Webs of Caring Relations

Cultivating Ecorelational Literacies in the Environmental Humanities

13 Wastelands of Decolonial Resurgence

Managing Land and Refuse/al in Gerald Vizenor's and Leanne Betasamosake Simpson's Indigenous Narratives

Elpida Ziavra

This chapter engages the Indigenous *undercommons*[1] of decolonial resurgence and love in Gerald Vizenor's *Earthdivers: Tribal Narratives on Mixed Descent* (1981) and *Landfill Meditation: Crossblood Stories* (1991), as well as in Leanne Betasamosake Simpson's *Noopiming: The Cure for White Ladies* (2022). The texts react critically to the mismanagement of waste by settlers; reject the capitalist overproduction and outsourcing of garbage to areas designated as wastelands; and affirm the Native commons' survival, care for, and relationality with land and other-than-human worlds. The Indigenous figures in the literary works under examination revalue and repurpose—in, at times, unexpected ways—polluting byproducts discarded by the North American consumerist culture. They create shelters, homes, reserves, artful objects, blankets, and shrouds and manage plastic so that this harmful material might be upcycled and not entirely wasted. The texts' aim is not to romanticize waste and depict a harmonious coexistence with toxicity, but to complicate who and what are theorized as dirt and which areas are managed as wastelands. This chapter also zooms in on the motif of the hands (*manus-i* in Latin) or paws that manage, manure, manufacture, build, grab specks of dirt and shape them into islands, hold, form reservations out of trash, sew, quilt, and upcycle waste into practical items or art. As Joshua T. Anderson beautifully argues in this volume, "[t]he word management reminds us that the power to change our story is in our hands."[2]

Vizenor's character Martin Bear Charme, recycled in both the *Earthdivers* and *Landfill Meditation*[3] stories, is a trickster figure and oral storyteller. Charme incarnates the author's speculative and ludic literary management of waste and, as an earthdiver, plunges into trash to create his wasteland of decolonial love[4] and communal healing. Vizenor coins the term earthdiver to refer to the animals which, in the Anishinaabe cosmogonic story, dove into the bottom of the flooded earth to grasp and bring to the surface grains of dirt in order to build Turtle Island, a solid ground on which to dwell. In the selected stories, Charme manufactures "his own turtle island,"[5] that is, a

DOI: 10.4324/9781003581185-17

reservation made out of garbage, which turns into a center for meditation and teaching among trash and foul smells. Charme's garbage island challenges the binaries of clean versus dirty, landfill versus protected area, and wasteland versus home and invites people to reimagine and re-story[6] their connection to trash. The shaman plays with literal and metaphorical waste and creates a theoretical framework that speaks to decolonial ways of managing land and refuse. At the same time, he attempts to fertilize the thoughts of his students and audiences with an ethics of care for the *crossblood*[7] human and more-than-human others, even with what is commonly considered as trash. Simpson's hybrid novel and poetry collection *Noopiming: The Cure for White Ladies* bears eclectic affinities with the theoretical framework of re-signifying trash that Charme preaches about. The text counters Susanna Moodie's 1852 colonial narrative[8] that erased Indigenous presences *in the bush* (the English translation of the word *noopiming*), a place settlers saw as savage, a *terra nullius* available for exploitation and colonization. *Noopiming* is haunted by the pollution of lakes and rivers and the precarity of plants and animals in the urban environment, threatened with extinction and loss of habitat. It is also critical of consumerist culture, which produces exorbitant quantities of plastic, especially polyethylene, inundating the world and penetrating everywhere in the form of microplastics. And yet, just like Vizenor's trickster narratives, *Noopiming* is not immersed in despair and lamentations over the ecological catastrophes that continue to dispossess Indigenous communities in twenty-first century North American metropolises (and elsewhere). Instead, the hybrid text underscores, with humor and irony, the "affirmative refusal"[9] of Native communities, thus enacting what Simpson rightfully argues is "an appropriate response to oppression, and within this context it is always generative; that is, it is always the living alternative."[10]

Vizenor's and Simpson's speculative texts also resonate with real-life, contemporary Indigenous communal strategies of refusal and reclamation of land,[11] as manifested by Simpson in her talk "A Short History of the Blockade." Simpson's orature foregrounds Indigenous ways of building relationality through dancing, singing, ceremony, language learning, storytelling, and blockades, which constitute "a refusal of the dominant political and economic systems of Canada."[12] She narrates the resistance of the Wet'suwet'en people against the settler management of the Coastal GasLink pipeline, which would pass through Indigenous lands, without the people's consent. The community fought back with a dissemination of their message and world-making through "videos, a website, and camp tours, and they have built the alternative—a land immersive community with cabins, a pit house, bunkhouses, and a healing centre at the Unist'ot'en camp."[13] The communal creation of a "Wet'suwet'en world, where no means no, with Indigenous bodies and Indigenous land,"[14] evokes the otherwise[15] worlds imagined and built with Charme's and the *Noopiming undercommons*'[16] decolonial efforts, love, and care. It also resonates with Audra Simpson's analysis of refusal as a politics, as well as "a symptom, a practice, a possibility for doing things differently, for thinking beyond the recognition paradigm"[17] and for identifying Indigenous forms of governmentality that defy the settler state and its organization. The texts under study unravel, therefore, the potential of decolonial narratives and artworks, which reverberate with an agonistic politics, to voice their grievances against environmental racist practices of land management like the dams that violently interrupt waterways and flood the surrounding areas or the mines, oil drill sites, and pipelines that pierce and scar the land.

Wasteland Meditations

Vizenor's selected postmodern stories, brilliantly crafted through multiple layers of embedded narratives in which the same characters and stories reappear, problematize the politics of identifying certain elements (and people) as dirt, excess, and waste to be discarded in lands marked as pollutable landfills. Vizenor's earthdiver, Bear Charme, dives into landfill waste to create his *postindian*[18] decolonial reservation of communal healing, where people are invited to think with trash "on the other end of the wasted world."[19] Charme's life story is unconventional: He was initially trained at welding, "[b]ut scrap connections bored him, so he turned his attention to scavenging and made his fortune hauling trash and filling wetlands with urban swill and solid waste."[20] The shaman earned large amounts of money by turning wetlands into wastelands and removing trash from the urban centers to natural areas, where it would not be seen or smelled by the sensitive noses of city-dwellers. However, Charme changed his career once again through an absurd, surreal move; he repurposed one of these swill and waste-filled areas into his refuge of peace and love and transformed landfill waste into the material for building his center of communal healing. Charme's land, considered a wasteland at first, turned into a valuable asset after the shaman constructed his unconventional sanctuary on it:

> His meditation reservation was once a worthless mud flat that he bought with a federal loan and covered with waste. Now his lush refuse reservation on South San Francisco Bay near Mountain View is worth millions. Charme has petitioned the federal government for recognition as a sovereign meditation nation."[21]

The expression "worthless mud flat" is a highly ironic critique of what Traci Brynne Voyles calls *wastelanding*, that is, a conscious strategy by settlers, who select areas they perceive as empty (and are usually inhabited by or are situated very close to BIPOC communities)[22] and without value, as they do not actively serve capitalism, in order to employ them as dumping grounds or camps for nuclear tests. *Wastelanding* rhetorically and materially constructs wastelands based on environmental racism and the systematic classification and hierarchization of lands and human beings as more or less worthy of being protected from pollution, toxicity, and their insidious and sometimes latent consequences.[23] Voyles argues that "colonial epistemologies do not just look on deserts as wastelands but . . . wastelands of many kinds are constituted through racial and spatial politics that render certain bodies and landscapes pollutable."[24]

Charme's *tricksterese* discourse allows him to speak at two different levels simultaneously; he refers both to literal trash, which he has reconfigured to build his reservation, and trash as a metaphor that dramatizes the (neo)colonial strategy of classification and categorization of certain racialized or lower-income peoples as dirty and less-than-human. *Wastelanding* has also been historically employed for the deracination and dispossession of Indigenous tribes from their lands, sending them off, like abject dirt, in arid areas unfit for human habitation and cultivation. As Mary Douglas argues, "[d]irt is the by-product of a systematic ordering and classification of matter, in so far as ordering involves rejecting inappropriate elements."[25] The landfill shaman reminisces how Indigenous peoples and their cultures have been systematically described and treated as parasites or unwanted "matter out of place:"[26] "[O]n the reservations the tribes were the refuse. We were the waste, solid and swill on the run, telling stories from a discarded culture to amuse the colonial refusers."[27] Charme playfully fuses the heteronymy of the noun *refuse* signifying trash and the verb *to refuse,* in order to demonstrate the dispossession and abuse suffered by Indigenous peoples in the hands of the settlers, or "colonial refusers," who disregarded and denied the humanity and rich cosmologies and cultures of Indigenous communities. A "refuse reservation,"[28] thus, signifies both a reservation made of waste and a space where Native and *crossblood* people not only refuse, in their turn, their refusers but also re-story their humanity and their conviviality[29] with other cultures and with all their kin. As Jennifer Gillan argues:

> The interdependence of cultures and species is important to Vizenor. Out of these exchanges comes growth. Thus, although reservations often have been dumping grounds for people as well as for technological wastes, they also have been compost heaps out of which new growth has sprung.[30]

The fermentation of waste, transformed into highly useful fertilizer, and the upcycling of certain elements commonly thrown away as trash resonate with Charme's reconceptualization of the *wastelanded* communities as convivial spaces of "affirmative refusal," decolonization, and intercultural cross-fertilization.

In Charme's landfill reservation, the meditators also speculate on *postindian* theories and stories that disregard the idealized and romanticized stereotype of the *indian* and attempt to see a fuller scope of the Native experience. Charme rejects *indian* exceptionalism and refuses to passively play the role of either the natural man who inherently protects Mother Nature or the perennial poor victim in need of rescue by settlers. As David Murray argues, Charme's landfill reservation aims to "mount an attack on notions of purity and geopiety, in which the earth and Nature are used as sources of spiritual value uniquely accessible to Indians."[31] The discourses which romanticize indigeneity as a way of being that is innately closer to nature do not reflect the complex, traditional Indigenous systems that ask for the constant work of creating and respecting the relational networks and treaties with the rest of creation.[32] These complex cultural and cosmological systems of being-with resonate with what Donna Haraway calls a *sympoietic*[33] effort to build, manage, and maintain non-taxonomical, multispecies worldings that respect and care for all kin. If Indigenous communities break the treaties and disregard their relationality with the rest of their kin, they are neither exempt from corruption and the capitalist logic of profit (Charme's scavenging years are a prime example of that) nor innocent of polluting the earth.

Charme underscores the hypocrisy of people who take pride in their ability to keep their houses clean while they forget the garbage they dump in their *wastelanded* backyards;[34] he urges his listeners via chant to fuse with the status of refuse, as the more people try to repress it, the sicker they will become: "Charme chanted *come to the landfill and focus on real waste*" [. . .] '*Mandala mulch, and transcend the grammatical word rivers, clean talk, and terminal creeds*, and put the mind back to earth. Dive back to the earth, *come backward to meditate on trash* and swill and *real* waste that binds us to our bodies and the earth (emphasis original)'.[35] As Charme emphasizes in the chant, humans are a part of the earth, producing waste and inevitably becoming organic waste after death, when they return to the soil to manure it;[36] thinking that they are above other creatures or that they can be absolutely pure, white, and clean is an illusion or a sign bordering on derangement.[37] Charme asks his listeners to resist the clichés that trap them in stereotypical misrepresentations by continuing to narrate their *postindian* stories of refusal that fight environmental injustices through humor, laughter, and irony. Charme's theory promotes conviviality in a multispecies, *sympoietic* commons. It aims at deconstructing the dominant ontological and political grammars and at decolonizing his students' and audiences' minds, by unsettling the arbitrary labels of (racial) purity and cleanliness. Charme performs a re-storying of the human garbage relations, as, as Haraway argues, "[i]t matters what stories make worlds, what worlds make stories."[38] Shifting the viewpoint of our stories, human beings can develop an environmental ethics based on care, even for the ones commonly conceived as the abject, refused, and rejected parts of the system.

Plasticity in a Plastic City

Noopiming is "a web or a network or an ecology"[39] and a "biosphere of Nishnaabeg caring"[40] that fleshes out a series of community-based praxes of resurgence and refusal to be and to act according to settler paradigms. The

Noopiming "multispecies commons,"[41]—consisting of two Elders, Akiwenzii and Mindimooyenh; a gentle giant named Sabe; a caribou named Adik; two young humans, Lucy and Asin; a maple tree, Ninaatig; and Mashkawaji, a hybrid creature and the *kwe*[42] narrator of the text, trapped at the bottom of a frozen lake—work together to protect, respect, and honor each other, as well as to counter the environmental racism weighing disproportionately on Native American lands and ways of being. Drowning in urban "plastispheres,"[43] the fictional figures demonstrate their plasticity and adaptability against a dominant culture of landfills that can be seen from space by reusing some of the discarded trash to make tents, refuges, and shelters necessary for their survival. They also create artistic interventions; when life gives them polyethylene, they make plastic art and tarp stars and voice their grievances against environmental disasters. In other words, they convivially "stay with the trouble"[44] and with each other, not in the utopian hope of returning to the times before settlers colonized the land known as Tkaronto, from which the modern-day name Toronto stems, but in achieving the political aim of homemaking and decolonial resurgence through care and love for their human and other-than-human kin.

The commons display an unorthodox relationship with trash, akin to Charme's methodology of meditating on and re-storying waste, in order to deconstruct the settlers' *wastelanding* policies that cast peoples and lands as pollutable and disposable. Akiwenzii, an Elder who carries the *gnoseologies*[45] of traditional ceremonies and of managing land through a deep engagement with it, is highly concerned with the uncanny omnipresence of polyethylene. However, they[46] refuse to tolerate pollution and attempt to bequeath their testimony of resistance and resilience to future generations. Akiwenzii travels to the Kinomagewapkong (Teaching Rocks) of their homeland to reflect, chant, pray, and meticulously carve, with their hands, a petroglyph of a petrochemical symbol that will remain long after their death as a potent reminder and a warning. The narration of Akiwenzii's rock-writing process follows a peculiarly slow rhythm, which records their day-to-day progress in a detailed manner and repeats multiple times, like a meditative chant, the Elder's nighttime carving on the rocks. The repetition creates anticipation; the readers follow the emergence of the symbols one by one by one till, slowly, a chemical compound starts to appear. First, it is C_2; then, it becomes C_2H; C_2H_4; and, finally $(C_2H_4)n$, the chemical symbol for polyethylene. Akiwenzii's unconventional petroglyph questions and refuses the settler strategies of actively and consciously transforming some sites into wastelands, while managing[47] other spaces and lands as sacred and untouchable, usually for reasons of tourism.[48] Akiwenzii does not

> need to try and explain that one can't just look at or preserve a sacred site. That if the sacredness is to be maintained, Nishnaabeg have to continue the relationship. Fast. Pray. Sing. Carve. You cannot just ignore something and expect it to still be there for you when you need it.[49]

The Elder performs a wake on the rock and creates a *stele* of sorts, so as to not only mourn but also testify to the continuous, hands-on love and care required in managing the land, manufacturing, and maintaining the relationality with its rocks, lakes, and rivers, contaminated by dangerous petrochemical substances like $(C_2H_4)n$.

Mindimooyenh,[50] another Elder of the community, is also concerned with the omnipresence and dangers of polyethylene overproduction. They say "(quoting Wikipedia): 'As of 2017, over 100 million tonnes of polyethylene are produced annually, accounting for 34% of the total plastics market.'"[51] Mindimooyenh vocally refuses the settler politics of excessive consumerism and waste mismanagement and gives their own meaning and value to what other people treat as trash. The Elder buys fruit and vegetables from the stacks marked as rubbish; they also collect, save, and resourcefully repurpose bits and pieces conventionally thrown away after one use, like toothpicks, fishing line, milk bags, scraps of paper, and used coupons. But their obsession is the multiple creative and innovative ways in which tarpaulin can be reused. The following excerpt demonstrates not only Simpson's playful and ironic tone toward consumerist culture but also the Elder's defiance and irreverence:

> Mindimooyenh is always waiting for the tarps to go on sale at Canadian Tire. . . . They put twenty 15×20 blue tarps in their cart and pay with old-school Canadian Tire Money with a tax rebate for their Status card. Mindimooyenh gives the lineup they created the finger on the way out because they will be making things count right up until the very end.
>
> Tarp as tent. Tarp as sleeping bag. Tarp as blanket.[52]

The Elder knows that the material they are so fond of because of its practicality and impermeability is not innocent, so they try to use it wisely and recycle it as much as possible. But Mindimooyenh does not employ the tarp only for their own personal needs; they try to help other people in various ways as well, one of which is to provide them with materials to create shelters and be protected from the elements. Mindimooyenh

> leaves a pile of Certified Value Tarps 15×20 in royal blue for whoever needs one. . . . [I]f someone new shows up without shelter, the Nishnaabeg throw one of Mindimooyenh's tarps over the plastic water bottle sculpture and call it "the Airbnb."[53]

—a good name for a shelter in the open air. The satirical tone of this excerpt underscores an otherwise worldview of refusal to give into a position of victimization, even when one is homeless, as well as the mockery of the settler habits of private property and overtourism.

The plastic bottle sculpture that serves as the skeleton for the Nishnaabeg "Airbnb" is one of the surreal artworks created by Sabe, a gentle giant who reconfigures trash they find in people's recycling bins around the city to make creative interventions: "For a while Sabe collected old toilets and sinks from rich neighborhoods and they would set them up in random places on the reserve. At the beach. In old Amos's backyard. In the field. In the sugar bush."[54] Sabe, like Charme, moves the trash from the city, where it is undesirable, to other locations and constructs imaginative islands of innovation and decolonial love, where junk gains new meanings and creates new stories. Sabe's installations recall Marcel Duchamp's controversial *Fountain*,[55] a readymade sculpture figuring a reversed porcelain urinal, an everyday object that became re-signified when it was exhibited as art. With their conceptual artworks, Sabe manages to disrupt the supposedly clear-cut distinction between an art gallery or museum and "wilderness," and they reappropriate certain spaces for the sake of their artistic freedom and expression. The giant's latest obsession consists of collecting small plastic bottles, which are stacked under a tarpaulin that "smells like cat pee and Sabe found it in a blue bin in the Avenues. As if you can recycle polyethylene."[56] And yet Sabe *does* recycle it. They take it from the trash and employ it as a protective tent for their plastic bottle sculpture, which will function as a structural framework for a shelter or a lodge. For the bottle sculpture, Sabe thinks of sewing "the noses of six bottles together into a star formation . . . and then sew all the stars together into a domed lodge."[57] This star-shaped design is purely decorative, as "six-pointed stars don't carry a lot of weight in Nishnaabeg cosmology."[58] In fact, Sabe "could make many more structures by simply making curved poles out of the water bottles, more like the saplings you'd use to make a sweat."[59] But Sabe allows themselves to be creative and unconventional; they refuse to treat all refuse as useless, and they find ways to create objects that will be beneficial not only for survivability but also for "Nishnaabeg joy"[60] and recreation.

After completing their unusual petroglyph, Akiwenzii dies on the Teaching Rocks, and their community embarks on an expedition to perform the death rituals for the Elder. Akiwenzii is shrouded in a quilt created by two young members of the community, Lucy and Asin, with tarpaulin stars stitched into it: tarp as blanket. Tarp as shroud. The blanket that embraces the Elder's body resonates with the clay sculpture *Trace* (2014)[61] by the Anishinaabe Canadian multidisciplinary artist Rebecca Belmore, a real-life *postindian* warrior and earthdiver. *Trace* speaks to the precarity of Indigenous survival in the urban "wastelands" of Canada and the potentiality of repurposing cheap, discarded materials into testimonies of resurgence and decolonial fugitivity. This skeletal structure hauntingly hangs over the visitors' heads in the Canadian Museum for Human Rights in Winnipeg, as a reminder not only of genocidal violence against Indigenous peoples and their lands but also of their perseverance and resilience. The artist created the monumental ceramic sculpture made out of hundreds of clay beads and shaped like a cape or a blanket full of holes, with the aid of her community.

For the manufacturing of such an immense artwork, a number of hands held and squeezed the soil collected from the land onto which the museum was to be erected, which "bears evidence of over 6,000 years of Indigenous presence where 500,000 artifacts were excavated."[62] The land contains and keeps the material traces and remains of the Native American bodies that lived and died on it, till they became compost that nourished and regenerated the soil. Each shape/shard is thus unique, the imprint of the palm and the fingers that carefully and lovingly touched and formed the clay into a lump that was pierced to become a bead and fired to harden up. The threaded beads were shaped into an immense blanket, a trace which conjures the traumatic history of smallpox blankets and biological warfare against Indigenous communities, deemed as parasites, dirt, and waste to be rejected. The blanket sculpture remains as a tangible symbol of traditional Indigenous cultures' resourceful love and care, of comfort and pride, even when made out of humble materials like Lucy and Asin's tarpaulin-star blanket. The interweaving of Vizenor's and Simpson's textual narratives of resurgence with the textile-like sculpture denotes the potential of fictional stories to be transformed from abstract metaphors into material, textural artifacts that speak to real-life events and resistances.

After Akiwenzii's wake and funeral rites, the *Noopiming* group also need to recover the body of Mashkawaji from the bottom of the frozen lake, where the ice is beginning to thaw. Hints of the frozen creature's rescue and reemergence appear at the laconic yet lyrical fragments in the poetic section of *Noopiming,* called "Mashkawaji's Theory of Ice."[63] The poem *weather* illuminates the agonizing moments during which Mashkawaji attempts, as an earthdiver, to grasp some dirt with the hope of building a new world. At those desperate moments, the hands of the Indigenous "constellation of co-resistance"[64] reach out to touch, grip, and lift the frozen being back to the surface:

swimming
drowning
sinking
i reach out
for my pawful of dirt
and instead
you hold my hand.[65]

Through the hybrid text, Mashkawaji narrates their "little stories feeding a network of connections that nourishes an affirmative Indigenous world, propelled by Nishnaabeg brilliance"[66] and their co-resurgence with plant, animal, human, and mineral kin, with spirits, shadows, and ancestors. Before they ended up in the lake, Mashkawaji was probably a warrior for environmental justice and care, who reacted against the pollution of the lakes. They fought against the poisoning that makes the water dangerous and jeopardizes the survival of all organic and inorganic kin:

> I'm thinking of the lake again. And how government scientists use a contraption to collect sediments from the bottom of a lake called an Ekman grab. It is a metal box on a string, with claws on the end. You trip it and the claws close, taking a sample of the bottom.[67]

This excerpt symptomatically reveals a parallel in the collection of samples by scientists with the earthdivers of the traditional Indigenous narrative; instead of the muskrat's organic paw, however, it is the cold, mechanical claw that reaches all the way to the bottom of the lake to grab its poisoned sediments, trap them into plastic bags, and send them to the lab: "The results always come back the same: they were right and there is nothing to be done because because. There is an important difference between testing and caring."[68] This metallic earthdiver-contraption reemerges to testify not to the creation of a new world, a solid ground for all creation, but to a tragic story of ecological catastrophes, irreversible damage, and lack of care.

Islands of Decolonial Resurgence

The texts under study portray otherwise world-building modalities, arising from communities residing in or near the areas marked as wastelands through a politics of radical, "affirmative refusal" and decolonial resurgence. The represented commons manage and reconfigure dirt and trash to narrate their histories and stories of creativity and innovation, reflecting on actual land reclamation efforts on Turtle Island and beyond, while inspiring, in their turn, future endeavors and actions. Vizenor's and Simpson's narratives affirm the sometimes quiet yet resistant modalities of being with organic and inorganic others or what Katarina O'Briain calls "quieter forms of Indigenous refusal that have always been there and have often been overlooked."[69] The earthdivers of the literary texts plunge their hands into trash bins, landfills, or dirt to manufacture and maintain their own turtle islands of decolonial resurgence, conviviality, and refusal. Various hands (fleshly or mechanical) and paws referred to in this chapter managed to clutch and hold lowly materials in their clenched fists: the earthdiving muskrat in the Anishinaabe cosmogonical story grasped and held on to two grains of dirt; Bear Charme hauled swill and garbage; the Ekman claw grabbed polluted lake sediments; the speculative commons of *Noopiming* recovered Mashkawaji and treasures from recycling bins; while Belmore and her community grabbed dirt full of ancestral traces. The various earthdivers then manifested their diverse ways of handling their findings: the trickster Nanabohzo built an entire island to protect and offer hospitality and love to all their relations; Charme built a hospitable refuse reservation for meditation and healing; the *Noopiming* characters constructed homes, boats, lodges, shrouds; and Belmore's commons created a monumental artistic exhibit in the form of a blanket.

In an era of fierce neoliberalism when large parts of humanity have been conditioned to overconsume without respect or concern for the consequences of their actions, Charme and his landfill reservation as well as *Noopiming*'s commons assume the heavy responsibility to re-signify and repurpose trash, acknowledging that waste and dirt are not separated from the human body, but are rather a central part of it. They know that absolute cleanliness is a dangerous illusion as it requires places to outsource and store waste, hands to manage and carry the refuse, and people who are deemed unclean and disposable enough to live in close proximity to it. The urban *postindian* warriors in these fictional works build their worlds, narrate, and inscribe their own histories and stories of *survivance*[70]—a term by Vizenor that signifies the continuation of storytelling, artmaking, and Indigenous traditions—by meditating on the arbitrary construction of wastelands with irony and playfulness. The characters refuse the refusers that categorize certain human beings as waste, pollutants, and dirty parasites that should be removed to vanish slowly somewhere else, in the wastelands of this world. This refusal is not a negative attitude that leads to stasis but an "affirmative refusal, resulting in the creation of not just points of disruption but collective constellations of disruption, interrogation, decolonial love, and profound embodiments of nation-based Indigeneity."[71] The texts under study reimagine, thus, the ways in which "[e]arthdivers, tricksters, shamans, poets, dream back the earth"[72] and dare to speculate on otherwise resurgent worlds, not as a dream for the future, but, as a fugitive decolonial potentiality, "an elsewhere in the here and now."[73] Refusing the (neo)colonial frames of thinking and categorizing humans, lands, and materials as trash allows for a generative rethinking and re-storying of a multispecies *undercommons*[74] where tricksters, landfill meditators, clay blankets, giants, Elders, rocks, trees, and all organic and inorganic entities *sympoietically* create and manage our complicated, messy, and convivial network of coexisting beings and intertwined worlds.

Notes

1 Harney and Moten, *The Undercommons*.
2 Anderson, "Badlands Management," pg. 20.
3 Charme appears in *Earthdivers: Tribal Narratives on Mixed Descent* (1981), *Landfill Meditation: Crossblood Stories* (1991), *Dead Voices: Natural Agonies in the New World* (1994), and *Chancers: A Novel* (2000).
4 This echoes with Simpson's book *Islands of Decolonial Love: Stories & Songs*.
5 Vizenor, *Earthdivers*, xv.
6 Taking my cue from Mayako Murai's introduction: "Re-Storying the World for Multispecies Survival," 1–5, I employ re-storying as a potent methodology for decentering, decolonizing, and rewriting the dominant discourses that disregard multispecies networks and relationalities.
7 This is Vizenor's term for a person with more than one ethnic identity.
8 See Moodie, *Roughing It in the Bush, or, Life in Canada*.
9 Simpson, *As We Have Always Done*, 198.
10 Simpson, *As We Have Always Done*, 33.
11 See Grover's keyword chapter on "Reclamation" in this volume for a historical overview of the term in the North American context and the reappropriation of the term by Indigenous communities, to voice their pleas for getting their land back.

12 Simpson, *A Short History of the Blockade*, 9.
13 Simpson, *A Short History of the Blockade*, 7.
14 Simpson, *A Short History of the Blockade*, 7.
15 I follow Daniel Heath Justice's reminder that "We can't possibly live otherwise until we first imagine otherwise" (156) and the collection of essays by Tiffany Lethabo King, Jenell Navarro, and Andrea Smith, called *Otherwise Worlds*, which examines the otherwise ontologies and ways of living together that emerge at the intersections between Black and Indigenous communities. I also think with Catherine Walsh and Walter Mignolo, who expound on how decoloniality moves away from the "colonial matrix of power" (4) and its logocentrism, "constructing paths and praxis toward an otherwise of thinking, sensing, believing, doing, and living" (4).
16 Harney and Moten, *The Undercommons*.
17 Simpson, "The Ruse of Consent and the Anatomy of 'Refusal'," 12.
18 *Postindian* does not signify after the *indian* for Vizenor; it is a term that aims to deconstruct the absurdity of the empty signifier *indian* with irony and a ludic attitude. See Vizenor's, *Manifest Manners*, 1–44.
19 Vizenor, *Landfill Meditation*, 101.
20 Vizenor, *Landfill Meditation*, 100.
21 Vizenor, *Landfill Meditation*, 100–101.
22 See Mohai and Saha's article on environmental justice and the way commercial hazardous waste treatment, storage, and disposal facilities are disproportionately placed close to low-income or BIPOC communities.
23 See the case of Love Canal, an area that was employed for dumping toxic waste and then hosted a school and a low-income neighborhood. Only when people started getting sick did the scandal erupt.
24 Voyles, *Wastelanding*, 10.
25 Douglas, *Purity and Danger*, 36.
26 Douglas, *Purity and Danger*, 36.
27 Vizenor, *Landfill Meditation*, 101.
28 Vizenor, *Landfill Meditation*, 100.
29 For the term conviviality, I build from Maya Hey's brilliant reading of human-microbe relations, wherein she cultivates a "coming together of disparate parties that go beyond a mere tolerance of differences," 23–24.
30 Gillan, "Restoring the Flow," 248.
31 Murray, "Crossblood Strategies in the Writings of Gerald Vizenor," 223.
32 See Mitchell, *Revenant Ecologies*, especially pp. 12–18, for two examples of Indigenous stories that see the world as an intertwined system, where balance is achieved through relationality, respect, and treaties that require work and constant revisioning.
33 See Haraway, "Sympoiesis Is a Simple Word; It Means 'Making-with'," 58.
34 The concern with the cyclical nature of pollution that inevitably ends up plaguing the polluters resurfaces in Vizenor's, *Dead Voices*, 57–58, where a community of fleas survive the chemical genocide of the insecticide-spraying exterminator, bite him, and remind him that: "The fleas that sucked your blood were chemical mutants, the dead voices of your poison now run in your blood, and the vengeance of the fleas is only the beginning of the war with chemical civilization."
35 Vizenor, *Earthdivers*, 92, emphases in the original.
36 See Cadava, "The Guano of History"; Agamben, et al. *Cities without Citizens*, 137–165, for an insightful analysis of guano as a central political concern in nineteenth-century America. The desperate search for this resource underscores the abuse and racism against BIPOC and their destiny to become guano sooner rather than later and fertilize the American soil and its rampant expansionism.

37 The blindness toward the consequences of environmental pollution that inevitably haunt the polluters themselves resonates with an arrogant and irrational way of thinking that Amitav Ghosh calls the "great derangement."
38 Haraway, *Staying with the Trouble*, 12.
39 Ramayya, "'Rehearsal for the World-Building Outside of Colonialism'," 5.
40 Ramayya, "'Rehearsal for the World-Building Outside of Colonialism'," 51.
41 Baynes-Rock, "Life and Death in the Multispecies Commons," 210–227.
42 See Simpson, *As We Have Always Done*, 27–33: *Kwe* does not only signify women but also resonates with the unique way Anishinaabe people conceptualized gender not as a fixed identity, but as a fluid modality of being and a "resurgent method," (27) "[a]t its core . . . about refusal" (33).
43 Eriksen, "The Plastisphere—The Making of a Plasticized World," 153–163.
44 Haraway, *Staying with the Trouble*.
45 See Mignolo, *The Politics of Decolonial Investigations*, 24, where Mignolo argues for a decolonization of the notion of epistemology and defines *gnoseologies* as "the conditions of all knowing, not only disciplinary" Eurocentric knowledge.
46 All characters in Simpson's text are referred to using they/them pronouns.
47 Simpson ironically illustrates another example of settler land management of a Toronto park: "more recently the tree cops have taken an interest in 'managing' the forest. That means new bathrooms, new signs, charging a ransom to hike the trail, and pruning," 97.
48 See Balachander's keyword chapter on "Rural/Wilderness" in this volume for a definition of *wilderness*, which needs to be preserved (and emptied of its Indigenous inhabitants): "Wilderness requires both physical and full ideological removal of Indigenous people to reframe lands as 'untouched' recreational landscapes for settlers" pg. forthcoming.
49 Simpson, *Noopiming*, 147.
50 A name that not only signifies old woman but "it also means the one who holds everything together" (The Writers, 2020, 15:41).
51 Simpson, *Noopiming*, 126.
52 Simpson, *Noopiming*, 85.
53 Simpson, *Noopiming*, 99.
54 Simpson, *Noopiming*, 98.
55 See Roche, Wood, and Duchamp, *The Blind Man*, 4–6, for an overview of the refusal the readymade sculpture faced and Duchamp's vocal response.
56 Simpson, *Noopiming*, 98.
57 Simpson, *Noopiming*, 175.
58 Simpson, *Noopiming*, 175.
59 Simpson, *Noopiming*, 175.
60 Simpson, *Noopiming*, 175.
61 Belmore, *Trace, Canadian Museum for Human Rights*.
62 Martin, "Rebecca Belmore's *Trace*."
63 Simpson, *Noopiming*, 205–225.
64 Simpson, "Indigenous Resurgence and Co-Resistance," 27.
65 Simpson, "Indigenous Resurgence and Co-Resistance," 211–212.
66 Meland, "Noopiming," 184.
67 Simpson, *Noopiming*, 181.
68 Simpson, *Noopiming*, 181.
69 O'Briain, "Refusing Settler Georgics," 9.
70 Vizenor, *Survivance*.
71 Simpson, *As We Have Always Done*, 198.
72 Vizenor, *Earthdivers*, xvi.
73 Jarrett and Ritskes, "Fugitive Indigeneity," I–XII.
74 Harney and Moten, *The Undercommons*.

Bibliography

Agamben, Giorgio, Shūsaku Arakawa, Madeline Gins, and Branka Arsić. *Cities without Citizens*, edited by Eduardo Cadava and Aaron Levy. Slought Books, 2003.

Baynes-Rock, Marcus. "Life and Death in the Multispecies Commons." *Social Science Information* 52, no. 2 (2013): 210–227. https://doi.org/10.1177/0539018413477521.

Belmore, Rebecca. "*Trace*: Canadian Museum for Human Rights." Accessed January 5, 2025. https://www.rebeccabelmore.com/trace/.

Douglas, Mary. *Purity and Danger: An Analysis of the Concepts of Pollution and Taboo*. Routledge, 1992.

Eriksen, Marcus. "The Plastisphere: The Making of a Plasticized World." *Tulane Environmental Law Journal* 27, no. 2 (2014): 153–163. http://www.jstor.org/stable/43294161.

Ghosh, Amitav. *The Great Derangement: Climate Change and the Unthinkable*. The University of Chicago Press 2017.

Gillan, Jennifer. "Restoring the Flow: Comic Circulation in Gerald Vizenor's Fiction." *North Dakota Quarterly* 67, no. 3–4 (2000): 242–255.

Haraway, Donna Jeanne. *Staying with the Trouble: Making Kin in the Chthulucene*. Duke University Press, 2016.

Harney, Stefano, and Fred Moten. *The Undercommons: Fugitive Planning & Black Study*. Minor Compositions, 2013.

Hey, Maya. "Symbiosis and the Steward: Reading Human-Microbe Relationships and Restoring Convivial Futures." *Synthesis* 15 (2022): 6–28. https://ejournals.epublishing.ekt.gr/index.php/synthesis/article/view/35717.

Justice, Daniel Heath. *Why Indigenous Literatures Matter*. Wilfrid Laurier University Press, 2018.

King, Tiffany Lethabo, Jenell Navarro, and Andrea Smith, eds. *Otherwise Worlds: Against Settler Colonialism and Anti-Blackness*. Durham: Duke University Press, 2020.

Martin, Lee-Ann. "Rebecca Belmore's *Trace*: Hands of Generations Past and Those That Will Come." 2014. https://www.rebeccabelmore.com/trace/.

Martineau, Jarrett, and Eric Ritskes. "Fugitive Indigeneity: Reclaiming the Terrain of Decolonial Struggle through Indigenous Art." *Decolonization: Indigeneity, Education & Society* 3, no. 1: (2014): I–XII. https://jps.library.utoronto.ca/index.php/des/article/view/21320/17382.

Meland, Carter. "Noopiming: The Cure for White Ladies by Leanne Betasamosake Simpson (Review)." *Native American and Indigenous Studies* 9, no. 2 (September 2022): 183–184. https://doi.org/10.1353/nai.2022.a863613.

Mignolo, Walter. *The Politics of Decolonial Investigations*. Duke University Press, 2021.

Mignolo, Walter, and Catherine E. Walsh. *On Decoloniality: Concepts, Analytics, Praxis*. Duke University Press, 2018.

Mitchell, Audra. *Revenant Ecologies: Defying the Violence of Extinction and Conservation*. University of Minnesota Press, 2023.

Mohai, Paul, and Robin Saha. "Which Came First, People or Pollution? Assessing the Disparate Siting and Post-Siting Demographic Change Hypotheses of Environmental Injustice." *Environmental Research Letters* 10, no. 11 (2015): 115008. https://doi.org/10.1088/1748-9326/10/11/115008.

Moodie, Susanna. *Roughing It in the Bush, or, Life in Canada*. McClelland & Stewart, 2007.

Murray, David. "Crossblood Strategies in the Writings of Gerald Vizenor." *The Yearbook of English Studies* 24 (1994): 213–227. https://doi.org/10.2307/3507892.

Murai, Mayako. "Re-Storying the World for Multispecies Survival." *Synthesis* 15 (2022): 1–5. https://ejournals.epublishing.ekt.gr/index.php/synthesis/article/view/35716/26798.
O'Briain, Katarina. "Refusing Settler Georgics." *Eighteenth-Century Fiction* 36, no. 1 (2024): 7–36. https://doi.org/10.3138/ecf.36.1.7.
Ramayya, Nisha. "'Rehearsal for the World-Building Outside of Colonialism': A Conversation with Leanne Betasamosake Simpson and Billy-Ray Belcourt." *Wasafiri* 37, no. 3 (2022): 50–56. https://doi.org/10.1080/02690055.2022.2067199.
Revkin, Andrew C. "Love Canal and Its Mixed Legacy." *The New York Times*, November 25, 2013. https://www.nytimes.com/2013/11/25/booming/love-canal-and-its-mixed-legacy.html.
Roche, Henri-Pierre, Beatrice Wood, and Marcel Duchamp, eds. *The Blind Man*, vol. 2, 1917. http://sdrc.lib.uiowa.edu/dada/blindman/2/index.htm.
Simpson, Audra. "The Ruse of Consent and the Anatomy of 'Refusal': Cases from Indigenous North America and Australia." *Postcolonial Studies* 20, no. 1 (2017): 18–33. https://doi.org/10.1080/13688790.2017.1334283.
Simpson, Leanne Betasamosake. *As We Have Always Done: Indigenous Freedom through Radical Resistance*. Indigenous Americas. University of Minnesota Press, 2017.
Simpson, Leanne Betasamosake. "Indigenous Resurgence and Co-Resistance." *Critical Ethnic Studies* 2, no. 2 (2016): 19–34. https://doi.org/10.5749/jcritethnstud.2.2.0019.
Simpson, Leanne Betasamosake. *Islands of Decolonial Love: Stories & Songs*. Winnipeg: ARP Books, 2013.
Simpson, Leanne Betasamosake. *Noopiming: The Cure for White Ladies*. University of Minnesota Press, 2021.
Simpson, Leanne Betasamosake. *A Short History of the Blockade: Giant Beavers, Diplomacy, and Regeneration in Nishnaabewin*. CLC Kreisel Lecture Series. University of Alberta Press, 2021.
Vizenor, Gerald. *Chancers*. University of Oklahoma Press, 2000.
Vizenor, Gerald. *Dead Voices: Natural Agonies in the New World*. University of Oklahoma Press, 1994.
Vizenor, Gerald. *Earthdivers: Tribal Narratives on Mixed Descent*. University of Minnesota Press, 1981.
Vizenor, Gerald. *Landfill Meditation: Crossblood Stories*. Wesleyan University Press, published by University Press of New England, 1991.
Vizenor, Gerald. *Manifest Manners: Narratives on Postindian Survivance*. University of Nebraska Press, 2010.
Vizenor, Gerald, ed. *Survivance: Narratives of Native Presence*. University of Nebraska Press, 2008.
Voyles, Traci Brynne. *Wastelanding: Legacies of Uranium Mining in Navajo Country*. University of Minnesota Press, 2015.
The Writer's Festival. "Noopiming: The Cure for White Ladies with Leanne Betasamosake Simpson." *Youtube Video*, 15:40. September 20, 2020. https://www.youtube.com/watch?v=3C7MUuBFmoc.

14 Resistant Infrastructure

Relational Responses to Ecological Punishment in Contemporary Multiethnic Narrative Practices

Dominique Aurilla Vargas

To encounter a physical barrier is to experience a shift in both mental and physical processes. When a body in movement arrives at a boundary, there is at least a momentary pause. For humans, this often results in a kind of arrested development: executive functioning must grapple with what comes next, and so the body shudders, rocking back from the shock of interrupted momentum. Within the psyche, the human must also either repress or process—in the moment or after—what this barrier communicates about the social context in which the human exists. For the more-than-human, interactions with barriers vary, but usually result in slowing, pooling, or pacing. Tentative efforts toward overcoming the barrier result in varying states of alterity for the human and more-than-human. Alternative routes might form, barriers may be overcome, but the body moves.

Beginning with this phenomenological account reorients abstract concepts of borders and barriers to highlight encounters between the human and more-than-human world and the phenomena through which the encounters are constituted. Understanding phenomena and how they are experienced and made sense of in the everyday can more clearly illuminate a host of affective and ecological effects of exclusion, enacting sociality and reifying geopolitical boundaries. These political and social barriers, usually produced in tandem with the built environment and border infrastructure (walls, fences, floating barriers) to mimic natural boundaries, effectively naturalize division and exclusion while also expressing dominion over the more-than-human world. The particulars of these interactions may then reveal the covert processes embedded within social relations, rendering them visible.

The infrastructures that lie beneath everyday human interactions are innumerable; however, they are—by design—systems with which humans have few intentional, conscious, everyday relationships. This covert quality allows for both political and corporate manipulation of relationships between public and private as well as discursive control over the function of infrastructure, relying upon a paradoxical controlled visibility. Post-industrial critical infrastructures, designed to "go unnoticed" by the very industries that will eventually benefit from proposed privatization, only surface in public consciousness when they break down, are destroyed, or converted "into

DOI: 10.4324/9781003581185-18

a proper spectacle."[1] Even so, visible spectacles of infrastructure hold an important place in the US imaginary. Settler-colonial processes historically promote infrastructure as progress. In this way, rail and roadways, plumbing, electrical grids, dams, pipelines, etc., become networks that encourage and often represent sociality. The very structures that seem to disappear and function beneath the built environment become symbols, signaling American exceptionalism and triumph over nature in service of political and economic policy. This belief is articulated in John Gast's 1872 painting *American Progress*, in which successive waves of "progress" follow a central angelic figure who moves through the frontier laying telegraph wires and leading steam engines West. Gast's painting implicitly suggests that infrastructure literally brings light and drives away danger. In this painting, the danger is depicted as darkness, wild animals, and Indigenous peoples. As such, infrastructure also implies a coordinated, "planned violence."[2] Westward expansion is synonymous with infrastructures that manage land and natural resources while also solidifying racial hierarchies.

Moreover, the historical approaches to settler-colonial land management in service to infrastructure enshrine this dominion in ways that tax the resiliency of natural systems and create conditions that contribute to increasingly devastating natural disasters such as wildfires and ongoing social inequities through environmental racism. Attuned to ongoing negative outcomes, Congress enacted HR 3684 Infrastructure Investment and Jobs Act in 2021, prioritizing natural resources-related infrastructure, wildfire management, and voluntary ecosystem restoration. In 2024, federal agencies like the Bureau of Land Management (BLM) insisted that "effective management . . . relies on natural infrastructure" (sometimes called "green infrastructure") and humans must work "with the landscape: to restore critical ecosystems and deliver environmental, social and economic benefits."[3] However, these processes do not occur in a vacuum. Climate change is steeped in racial capitalism and gender-based violence, affecting the most vulnerable populations more severely. If the primary goal of any response to this complex problem is maintenance of the neoliberal status quo and economics, then these changes cannot begin to mitigate the damage done to human and nonhuman relations. Furthermore, even land management policies that seek to repair the damage of settler-colonial practices must be contextualized with the policies of other governmental agencies such as US Customs and Border Protection whose mandate requires integrated technologies including "infrastructure management" in the form of tactical, environmental intervention.[4]

US domestic and foreign policy in the Western hemisphere relies upon a simultaneous expansion and enforcement of national borders as well as intervention beyond them to articulate political and economic dominance. Following domestic Westward expansion, the Monroe Doctrine and the clarifying Clark Memorandum levied foreign policy to extend US power and influence across the Western hemisphere. Later, the creation of the North American Free Trade Agreement (NAFTA) as a "legal regime of trade and capital investment" resulted in the opening of "the borders among the three nations by changing the formal legal controls on the entry and exit of goods and capital, although notably not of people."[5] The signing of NAFTA is contemporaneous with border security operations like Hold the Line, Gatekeeper, and Blockade. Each of these military operations prioritized "Prevention Through Deterrence" in response to the anticipated increase in immigration to the United States due to NAFTA.[6]

The policies around US border maintenance require discursive control over the regulation of immigration, management of the natural environment, and facilitation of transnational economic policies. In response, US Customs and Border Protections (CBP) emphasizes "tactical infrastructure," which simultaneously engages environmental planning; building; and maintenance of "facilities, physical barriers, roads, lighting, surveillance systems, and detection technology." While CBP touts a commitment to "environmental and cultural stewardship," it does so in tandem with "core missions of border security and the facilitation of legitimate trade and travel."[7] While this mandate covers a variety of built barriers and surveillance technology, it cannot be read without the contemporary emphasis on a particular border barrier: the Wall. This barrier is itself a highly visible and storied monument to tactical infrastructure.

Border walls are generally understood to articulate state beliefs around purity by preserving the "integrity of those with established and legitimated status" and excluding others who threaten that cohesion.[8] The US Border Wall is only the most recent articulation of physical boundaries that have taken shape along the southern US border over decades of historical human deterrence efforts, but it is bound up in several affective valances. Militarization of the border is often justified as deterrence of undocumented immigration, but it requires elements of social and ecological control. Thus, the metaphors associated with border crossing connect the environmental "tide," "swell," and "flow" to the militaristic "invasion," justifying built barriers as both necessary and natural. Arguably, the goal of this barrier was never about deterrence alone; rather, the policies are meant to punish those who defy the border. The environment then becomes a necropolitical apparatus of border enforcement.[9] This is not to say that the environment is a willing participant in this punishment and surveillance. Rather, these discourses are mapped on to the more-than-human, which forcefully disrupts relationality by enforcing division between human and more-than-human beings.[10]

The infrastructural tools by which borders are enforced—walls, fences, marine barriers, and drones—often mimic natural barriers enacting political will on the landscape itself. This alteration and conscription of the natural environment beg the question of how the natural or more-than-human world is implicated in service to political ideologies. If we are to accept that borderspaces are often constituted through a "necropolitical ecology," we must consider how the environment is "brought into the service of such border maintenance," or if we might seek "less anthropocentric" ways of understanding the more-than-human. These questions require a nuanced approach that troubles the understanding of the environment as "purely passive" and "forced to carry out the death-work of humans in the state of exception."[11] As we consider moments of arrested movement, alterity, and relation, we must also seek to excavate oceans, rivers, and deserts from the infrastructure of punishment where they are at best passive or inactive participants and at worst servants to the regimes of deterrence and punishment.

It has long been the project of writers and artists who engage with border spaces to articulate the effects of infrastructures of punishment on the human and more-than-human world. However, this writing has not been historically placed within the tradition of nature writing. Lauret Savoy and Alison H. Deming write in their introduction to the anthology *Colors of Nature*: "Although writings about the natural world have existed from ancient times to the present in countless cultural traditions and languages," writers of color are often excluded from the Euro-American tradition of "nature writing" as a genre. This writing was predicated on "solitary explorations of wild places from a poetic, philosophical, or scientific perspective; seeing nature as a place apart," making nature writing by writers of color unrecognizable in its "intimacy with and response to the larger-than-human world." Savoy and Deming argue for "widening the frame" to describe this tradition expansively.[12] Responding to this call also requires recognition of the way that intimacy and relationality with the more-than-human may build a resistant infrastructure in defiance of the necropolitical ecology of punishment.

Building an archive of relational responses to the racial, gendered, and exclusionary discourses of border spaces is not difficult. Gloria Anzaldúa and Lauret Savoy both responded to the legacy of settler colonialism in racialized geography and contemporary geopolitical borders with multi-generic, ecologically-oriented writing. *Borderlands/La Frontera* and *Trace* share an aesthetic resistance to infrastructures of punishment in form and content, suggesting new ways of connecting with the more-than-human world: a decolonial environmentalism.[13] The poetry of Ofelia Zepeda and the performances of Las Nietas de Nonó further show that ecological writing may activate shared strategies of distributed selfhood and resistant worlding built on intimacy, trust, and love. Together, these texts resist the necropolitical appropriation of the more-than human world through embrace of alterity.

Geopolitical borders are often mapped over and through natural boundaries or spaces deemed "empty," such as rivers, oceans, and deserts, overwriting their ecological importance to the bioregions they sustain. The inscription and enforcement of barriers and borders can be ecologically and discursively devastating, revealing a pattern of implicating ecological features in violence. The Rio Grande, for example, maintains a fragile ecosystem and provides a vital water source to desert peoples, but it has been abstracted from the ecological to the political and etched onto the US-imaginary as the physical manifestation of the porosity of the US-Mexico geopolitical border. The metaphor exceeds even the political, becoming an affective and libidinal object.

In June 2023, as part of an ongoing dispute with the Biden Administration over border and immigration policies, Texas Governor Greg Abbott and the State Legislature enacted "Operation Lone Star," a state-funded effort to "secure the border; stop the smuggling of drugs, weapons, and people into Texas; and prevent, detect, and interdict transnational criminal behavior between ports of entry."[14] In the month following Governor Abbott's announcement of Operation Lone Star, the state's continued efforts to punish criminality and deter immigration reached a new phase. The Texas Department of Public Safety deployed a 1,000-foot-long string of 4-foot-high bright orange buoys that anchor to the riverbed and divide the Rio Grande. These buoys are designed to spin so that they impede climbing, but they also include serrated metal plates that ensure injury. Anchors underneath the buoys include webbing that inhibits swimming under them, and the US banks of the river are lined with multilayered razor wire. Prior to the deployment of this barrier, the river was already difficult to navigate and heavily patrolled. Though the US Justice Department filed suit against the state of Texas in 2023, demanding that the state remove the barrier, the Fifth Circuit Court of Appeals stayed a preliminary injunction. Following the 2024 US Presidential election, the use of floating barriers in the Rio Grande has expanded. Even though this barrier currently spans only a fraction of the 1,254-mile river, the state continues to insist that the buoys are necessary to secure the border. In this way, the natural boundary of the river is co-opted into the infrastructure of punishment and intimidation where the river remains a symbolic threshold abstracted from its own ecological processes.

Efforts to weaponize the natural environment to deter or punish immigration are not new. The border wall between Tijuana and San Diego is another important example of ecological punishment wherein the Pacific Ocean, marine life, and water currents become the apparatus of deterrence and punishment. Here, the interaction of the built and natural environment creates the ideal conditions for drowning and death. The wall at this point in the border juts out almost 200 feet into the ocean, where the metal posts from the wall create a permanent rip current that pulls swimmers out into the ocean. Avoiding dangerous swimming conditions leaves only one other option: to hold on to the wall and walk along a precarious ledge of shellfish and barnacles growing up the beams. The shells are sharp and slick, often resulting in severe lacerations. While these conditions have always been dangerous, an increase in the height of the wall in 2019 correlates to increased numbers of injuries and death.[15] The waters of both the river and the ocean are two of the few options available to would-be immigrants who are diverted from traditionally connected and "populous areas and into more remote and ecologically fragile parts of the border."[16] Those who attempt to avoid border walls, marine barriers, and militarized checkpoints are left with only one other choice: crossing the Sonoran Desert (for which they are ill-prepared), where they face almost-certain heat-related illness and death.

The Sonoran Desert spans the US-Mexico border, creating a distinct bioregion, uniting a landscape cleaved by the violence of geopolitical conflict, settler-colonial control, and shifting anxieties around the nation-state. The desert landscape is commonly assumed to be temporally fixed, peripheral, and hostile to human life.[17] As such, much of desert discourse focuses on the adversity and dangers of desert spaces. According to Natalie Koch, this discourse is deeply tied to US investment in the cultivation and transformation of so-called arid land, which wrested control of water and natural resources from Indigenous communities.[18] This outlook is exacerbated by state-sanctioned violence and militarized enforcement of the US-Mexico Border, so that the bioregion surrounding this geopolitical boundary has become implicated in its goals.

Anthropologist Jason de León theorizes a "tactical relationship between federal border enforcement policy" and the "harsh landscape" of the desert.[19] This relationship is shaped by settler-colonial policies, racial exclusion, and Prevention Through Deterrence policies, reframing the desert as a partner in deterrence. The desert is undoubtedly weaponized by the militarization of the border, but it cannot be argued that the desert (as a space and a more-than-human relation) consents to or actively participates in the violent maintenance of criminalized crossing. De León argues that the rhetorical strategies of Prevention Through Deterrence, which rely on the "agency of animals and other nonhumans to do the work of law enforcement," describe the desert as a "ruthless beast that law enforcement cannot be responsible for."[20] As such, state-sanctioned weaponization of desert space must be seen as part of a larger neoliberal effort to shift responsibility for death and injury to the individual (and the desert itself) rather than the state.

Anzaldúa's *Borderlands/La Frontera* offers a materialist history of the border along with a philosophical treatise on associated liminality. According to Lucinda Newns, Anzaldúa draws on natural features as "metaphors for the cultural and linguistic displacement felt by those straddling the 'herida abierta'. . . dividing the US from Mexico," offering "a model for reading the entanglements between human bordering practices and the non-human world."[21] Recent criticism of Anzaldúa's writing highlights, among other concerns, a problematic adherence to mestizaje and La Raza Cósmica. Newns also troubles Anzaldúa's tendency to naturalize the "relationship between Chicanas/os and the land," which she claims undermines the "environmentalist aspects of the text" and works against its "overall project of deconstructing essentialisms because of its inscription of land-based identities."[22] However, Anzaldúa's decolonial framework provides the tools through which we can critique even her tendency to lapse into essentialism. Critiquing infrastructures of punishment, Anzaldúa theorizes relation with the more-than-human, which exceeds the essentialist binaries of border logic.

Anzaldúa writes that borders are "set up to define the places that are safe and unsafe, to distinguish *us* from *them*" (emphasis original). In this way, the border demarcates an "undetermined place created by the emotional residue of an unnatural boundary,"[23] creating a "shock culture, a border culture, a third country, a closed country."[24] Yet, the river is not a space of division alone. Despite its appearance as a natural boundary and its representational quality, the banks of the river do not demarcate the end or beginning of a new environment. The Rio Grande is a watershed that connects both sides of the border as a bioregion or a "geographical area of similar climate where similar flora and fauna are found." Beginning in the 1970s, environmentalists suggested that bioregions represent a "form of communal consciousness, believing that citizen-like allegiances could be built according to bioregion in a way that challenges national boundaries."[25] For Anzaldúa, this communal consciousness is found in border culture formed by the bioregion of the Rio Grande as well as interaction with political borders. In its wholeness, the river's bioregion resists the imposition of geopolitical infrastructures.

Resistance to co-option under the infrastructure of punishment is palpable in *Borderlands*, emphasizing "nature's resistance" to "human-made boundaries" and the discursive power of settler colonialism.[26] Throughout *Borderlands*, water, wind, and land resist the unnaturalness of the geopolitical border. However, the descriptions of the land and water are also imbued with human engagement ("wind tugging at my sleeve") and intimacy ("earth touches ocean/where the two overlap/a gentle coming together") in addition to opposition ("at other times and places a violent clash . . . silver waves marbled with spume gashing a hole under the border fence").[27] The natural processes of the more-than-human exceed the powers of built infrastructure.

Anzaldúa's efforts to theorize ecological resistance require exposing the power of border discourses to shape human relationship to place. Similarly, Savoy argues that the process of "place-making" allows colonial forces to "possess territory on the ground," but, in so doing, "also lay claim to territory of the mind and memory of the future and the past."[28] In *Trace: Memory, History, Race, and the American Landscape,* Savoy makes visible the processes by which infrastructures of punishment covertly operate through geographic knowledge to further entrench settler-colonial racial hierarchies. In her reflections and memories of these places, Savoy often focuses on discrete details and landscape features through which she can analyze the past and present simultaneously, which ultimately blurs temporal as well as spatial boundaries.

Drawing on the landscape, Savoy builds a hermeneutic through which she is able to process experiences of exclusion. For example, while visiting the History Museum in Trinidad, Colorado, which she writes held "stunning portraits of African Americans and people of obviously mixed heritage" dating from 1889, Savoy expected to find information on the people in the photos or perhaps even her own ancestors who may have passed through the area. Nevertheless, the curators told her: "Little information is left on African Americans in this area—of course, they don't live here now. . . . Subject information for the . . . photos is almost nonexistent." She finds that many images are titled "Unidentified _____." She is troubled by the failure of archival power where images are "unidentified, unidentifiable," and Black US histories are erased. Savoy leaves the museum and realizes that though "people crowded the streets. No one looked like me." Erased in the past and present, Savoy goes on to describe her experience of this realization as "isolated grains of sand" teasing absent stories.[29] This granular quality of the archive provides a method through which Savoy may be able to understand the effects of epistemic violence on selfhood. In isolation, grains of sand may appear to be disparate or displaced, but by teasing absent stories, they also indicate that they exist in relation to what cannot be seen. For Savoy, the absent desert-worth of grains holds a distributed sense of self that may exist beyond what has been erased.

This alterity hints at the way that nondominant epistemologies resist settler-colonial erasure and ecological punishment by defining the self as a process of relation with the more-than-human world. Scholars like Margo Tamez (Lipan Apache) have described the way this relational process denies discursive control over land dispossession and Indigenous rights, which may build a resistant infrastructure outside of colonial methods. Tamez describes Ndé relationality as "deeply affected by borders, militarization and non-recognition;" nevertheless, "rivering," or the "dynamic knowing and being within rivered places," resists the way land was "barbed, gridded, and separated" by settlers, foregrounding connection and refusing dispossession.[30] Karen Barad similarly suggests that relational practices can alter discourse: "In an agential realist account, discursive practices are specific material (re)configurings of the world through which the determination of boundaries, properties, and meanings is differentially enacted."[31] These theories offer two accounts of the way relationality may resist singularity and control.

Encounters between human and the more-than-human then emerge as entanglements, exceeding the discursive control of colonial methods. While Anzaldúa and Savoy's environmentally engaged texts may seem generically incompatible with Ofelia Zepeda's poetry and Las Nietas de Nonó's engagement with SCOBY (symbiotic culture of bacteria and yeast), they share a critical focus on the inseparability of the self from the more-than-human world. Furthermore, they challenge the idea of a discrete individual altogether, calling for a pluralized conception of self. These texts model resistant intimacy through relation with the more-than human, which cannot be managed or impeded by unnatural borders or work in service to the infrastructure of punishment.

Importantly, this sense of a scattered selfhood does not indicate fragmentation. In *Trace*, the image of sand becomes a trope, illustrating not only liminality and the inarticulability of historical processes but also the power of accumulation. Framing *Trace* as re-membering, Savoy claims that the connection between the human and more-than-human is a shared wounding: "An immense land lies about us. . . . [T]he past looms close, as immediate as breath, blood, and scars on a wrist. It, too, lies hidden, obscured, shattered."[32] To uncover these patterns and entanglements, Savoy writes that she must "trace other residues of hardness." She settles on the transition from the Mojave Desert to the "Devil's Punchbowl," which serves as a "tactile reminder that here is a land of process and response."[33] Capturing the geological time of the mountains and desert, Savoy describes this area as a "geological trap door hidden by alluvium, freeways, and sprawl." Though Savoy is careful to show the entanglement of the human and more-than-human in this scene, she turns her focus back to the slow changes of the punchbowl, which "consists of sandstone and conglomerate, once sands and pebbles of ancient mountain streams that flowed millions of years ago, now upended into rocky tablets."[34] Foregrounding the slow, recursive movement of sediment seems to not only illustrate displacement and memory, but it also simulates the connection between mountains and desert. In addition to erosion and contributing silt, the San Gabriel mountains form alluvial fans that descend into the Mojave, creating groundwater reservoirs that sustain life in the desert. The natural infrastructure of sand and sediment also provides the space through which Savoy can filter her own experiences, resisting the violences of erasure and dispersal. To embrace the granular, dry, and shifting sands is to show love to an undervalued element and embrace a distributed selfhood that prioritizes relation over individualism, wholeness over fragmentation, and connection over isolation.

This affective connection between the human and more-than-human not only emphasizes resistance but also shows the flexibility of an infrastructural approach. Working alongside, engaging with, and showing reciprocity to the sands, rivers, oceans, winds, and more, the human subject is both healed and healing. These encounters are transformative, embracing alterity that rejects the status quo while imagining new modes. In the final chapter of *Borderlands*, Anzaldúa explores the power of affective excess. In this scene, border infrastructures seek to interrupt relationality, but the speaker, in the moment of encounter, overcomes these boundaries through intentional trust-building and sublimation of individual self. She begins with a description: "I stand at the river, watch the curving, twisting serpent, a serpent nailed to the fence where the mouth of the Rio Grande empties into the Gulf." While the image of the snake, a reference to the iconography of Aztlán and Coatlicue, is violently attached to a "fence" (the US-Mexico border), the river and the gulf interrupt this division. The speaker is enmeshed in the environment despite the division: "I shade my eyes and look up. The bone beak of a hawk slowly circling over me, checking me out as potential carrion. In its wake a little bird flickering its wings, swimming sporadically like a fish." Following the flight line of hawks (who she imagines view her as a potential source of nourishment) and smaller birds that move like fish in the water, she resists even accepted ecological boundaries of water and air.[35] In intimate communion with the more-than-human—the sky, the Rio Grande, the Gulf of Mexico, the land—the speaker experiences liquidity and connection. This mode of relation is not inaccessible to other racialized identities, just as it does not undo the material work of the border. However, drawing the line of sight away from the border and toward the sky, the speaker willingly allows herself to merge into the wholeness of the more-than-human. In this way, the infrastructural apparatus of punishment fails to inscribe the speaker with the totalizing necropolitical discourse of border enforcement.

Savoy's love and Anzaldúa's trust build a nascent sense of distributed selfhood. By sublimating the self and emphasizing connection, these texts oppose the infrastructure of punishment that serves to solidify racial hierarchies and enforces borders of temporality and nation. While built structures may reify the discourses of punishment, they are not capable of enforcing the borders of self. Resisting necropolitical geographic knowledge is then ultimately about embracing a distributed self in communion with more-than-human relations.

Ofelia Zepeda's poetry is marked by deep resonance with desert space and resistance to "the logic of U.S. and Mexican geopolitical divisions."[36] While this orientation reflects Tohono O'odham epistemologies, it also resists the infrastructures of punishment mapped on to Sonora by deterrence and enforcement strategies. Zepeda's poem "Birth Witness," from the collection *Where Clouds Are Formed*, narrates Zepeda's attempt to obtain a birth certificate, an experience precipitated by the militarization of the border. The poem begins with the story of Zepeda's birth. This story is explicitly linked to her mother's memory: "she remembers it was windy" and that the "tin roof rattled, a piece uplifted/from the wooden frame." The poem continues with an explicit link to place and time: she writes that her mother "knew it was March" and that she was "in the cotton fields of Arizona."[37] However, when she relays this story to the woman who stands in for the state, offering a baptismal certificate as proof of her birth location, it is rejected. She is told "it doesn't count," and she needs someone who witnessed her birth to vouch for her. Zepeda's poem proceeds to catalog myriad "silent witnesses" to her birth including the stars, wind, sun, and pollen. The more-than-human world offers comfort, presence, and silent reassurance in communion with her mother's labors. However, with her birth, Zepeda too becomes an integral part of this environment. She writes that with her first breath, she "took in those dry particles from the cotton fields;" in so doing, she becomes part of the land, and it becomes part of her.[38]

Though none of the elemental witnesses are acceptable to the state apparatus, they are the distributed particles that make up Zepeda's sense of self. To further emphasize this ongoing connection, Zepeda writes that her parents could not have made birth announcements or written her birth in a family bible because they are not literate in English; instead, they speak in relation with the earth. This language is "useful for pulling memory from the depths of the earth . . . praying with the earth and sky . . . singing songs that pull down clouds . . . calling rain." Zepeda insists that this sacred language is "too civil for writing minor things like my birth."[39] The alternative literacy that Zepeda describes is relational. Language reflects the way that self is understood as part of the more-than-human. While the poem provides justification for her birth story, Zepeda resists the state-sanctioned discourse that seeks to delimit her birth to verifiable facts that support the naturalization of the border. Replacing the language of management with elemental and ecological discourse, Zepeda draws on the distributed selfhood and relation with the more-than-human world to manage the story of her birth.

Zepeda's poetry builds this ecological/elemental literacy, exploring the Tohono O'odham relationship to water and rain in the Sonora. Zepeda further characterizes the desert not through land and aridity but through fluids and liquidity. She writes:

> Rain in the desert is a theme I continue in much of my writing. I don't claim that I write about the land. . . . I like to think that I am more influenced about the movements and changes in the atmosphere of the desert environment. Changes that occur most drastically when it rains.[40]

Like Anzaldúa's urge to look upward at the flightline of birds over the border, this insistence on atmospheric engagement speaks to the totality of human and more-than-human relation. This relationship cannot be contained to ground-level struggle with settler-colonial discursive emptiness and individualism.

Multiple valences of resistance promote an engagement with selfhood that extends beyond the singular. Embracing alterity, Las Nietas de Nonó, a creative collaboration between sisters mulowayi and mapenzi, explore this selfhood through performance and dance. Their early work with Patio Taller engaged directly with the exclusion of Black communities in Puerto Rico, which is attributable to the infrastructures of punishment in the colonial state: "racial and class discrimination, mass incarceration, drug trafficking, obstetric violence, and the cycle of poverty."[41] Since 2016, however, Las Nietas de Nonó have consistently engaged with SCOBY as both a method and medium in their performances. Leveraging connection to the landscape, these engagements "expose the neo-colonialist exploitation and over-industrialization threatening the environmental and human sustainability" of Puerto Rico.[42] In their installations, SCOBY products are alternately shaped, stretched, and woven into a variety of interactive materials, which demand sensory engagement. The artists have created screens out of this material onto which they projected images, but they have also worn garments fashioned from the SCOBY matrix in their performances.

Growing these cultures takes on familial terms: the SCOBY "mother" is "its own functioning community, a cellulose mat that reproduces a new 'baby' SCOBY."[43] While this interaction could be read as metaphors for intimacy, "interconnectedness and care,"[44] it is also a practice of hyperlocal relationality. Las Nietas de Nonó cultivate the SCOBY from bacteria and yeast, found and grown in the same location as their installations and performances.[45] In so doing, their engagement is different in each location, relating to the more-than-human world as dynamic and changing. Just as Anzaldúa insists that the human cannot be described as a single entity: "you're all the different organisms and parasites that live on your body and also the ones that live in symbiotic relationship to you . . . you're a multiple entity"[46]; Las Nietas de Nonó honor the multiplicity of distributed selfhood.

Las Nietas de Nonó's relationship with SCOBY reinforces connection and distributed selfhood of a resistant infrastructure. For Anzaldúa, this relationship supports a "spiritual belief . . . that we are indistinguishable from the environment."[47] Resisting the colonial and imperial structures that demand a singular sense of self, SCOBY cultivation and engagement embrace a self in dynamic relation to place and more-than-human, "demonstrating ways in which bodies, human and nonhuman, are intertwined."[48] Rather than treating SCOBY as mere material, this engagement represents a co-creative and resistant exploration of selfhood that enacts reciprocity.

Necropolitical ecology creates a state of exception in which the human and more-than-human are rendered vulnerable and violable. Furthermore, borders are everywhere, extending well beyond the physical geopolitical borders. Following the Homeland Security Act of 2002, Immigration and Customs Enforcement (ICE) was formed with the mandate to surveil and enforce border discourse across the nation. The infrastructure of punishment supports this mandate, extending ecological punishment to virtual realms, airports, schools, homes, and workplaces. In the age of disappearances, canceled visas, and expedited removals, the flexible technology of punishment extends border discourses to the individual. Resisting the infrastructure of punishment requires an engagement with alterity and reimagining of resistant infrastructure.

Barad suggests that the primary ontological unit is not independent (separate) objects but phenomena, which are "the ontological inseparability/entanglement of intra-acting agencies."[49] This is similar to Denise Ferreira da Silva's conception of the inseparability of matter, which asks:

> What if, instead of the Ordered World, we imaged each existant (human and more-than-human) not as separate forms relating through the mediation of forces, but rather as singular expressions of each and every other existant as well as of the entangled whole in/as they exist.[50]

Embracing a distributed self is a poetic, performative act that may result in alternate forms of knowledge and being. It is only through engagement with alternative infrastructures (particles, sand, moisture, bacteria) that the precariousness of this space comes to constitute a gap, peripheral and central, infinite and singular, where we can begin to experience relation with a space that is unsettled in every sense.

Notes

1 Rubenstein, Robbins, and Beal, "Infrastructuralism," 576.
2 Boehmer and Davies, *Planned Violence*, 107.

3 “Infrastructure.”
4 “U.S. Border Patrol Technology.”
5 Saldaña-Portillo, “In the Shadow of NAFTA,” 751–777.
6 De León, *The Land of Open Graves*, 5.
7 “Environmental Management.”
8 Casey, “Walling Racialized Bodies Out Border versus Boundary at La Frontera,” 190.
9 Mbembe, *Necropolitics*, 92.
10 For a full accounting of this disruption, see Lisa Fink and Lydia Cagluaq Agnus’s keyword chapter “Relationality” in this volume.
11 Newns, “Necropolitical Ecologies,” 41.
12 Deming and Savoy, “Widening the Frame,” 5–6.
13 As Rafael A. Martínez writes in his keyword chapter “(Un)documented Ecologies,” included in this volume, literature “represents an opportunity to document” the experiences of human and more-than-human world as they are shaped by the complex interaction of settler-colonial, man-made structures and the natural world.
14 “Operation Lone Star Boosts Border Response with New Marine Barriers.”
15 Lussier and Lindholm, “Rate of Aquatic and Maritime Drowning Deaths after US-Mexico Border Wall Height Increase,” 1059.
16 Newns, “ ‘The Sea Cannot Be Fenced’,” 1097.
17 For further examination of the representation of desert space, see Quinn Grover’s keyword chapter “Reclamation” in this volume.
18 Koch, *Arid empire*, 35.
19 León, *The Land of Open Graves*, 29.
20 De León, *The Land of Open Graves*, 43.
21 Newns, “ ‘The Sea Cannot Be Fenced’,” 1101.
22 Newns, “ ‘The Sea Cannot Be Fenced’,” 1108.
23 Anzaldúa, *Borderlands/La Frontera*, 25.
24 Anzaldúa, *Borderlands/La Frontera*, 33.
25 Newns, “ ‘The Sea Cannot Be Fenced’,” 1107.
26 Newns, “ ‘The Sea Cannot Be Fenced’,” 1106.
27 Anzaldúa, *Borderlands/La Frontera*, 23.
28 Savoy, *Trace*, 75.
29 Savoy, *Trace*, 22–23.
30 Tamez, “Indigenous Women’s Rivered Refusals in El Calaboz,” 10.
31 Barad, *Meeting the Universe Halfway*, 56.
32 Savoy, *Trace*, 29.
33 Savoy, *Trace*, 16.
34 Savoy, *Trace*, 16.
35 Anzaldúa, *Borderlands/La Frontera*, 111.
36 Huizar-Hernández, “Between Place and Plot: Reimagining the Story of Arizona,” 275.
37 Zepeda, “Birth Witness,” 12.
38 Zepeda, “Birth Witness,” 13.
39 Zepeda, “Birth Witness,” 13.
40 Zepeda, “Autobiography,” 412–413.
41 “Las Nietas de Nonó.”
42 Apostolatos, “Foraging the Fragile and Fertile Soil.”
43 Irving, “Artists-in-residence Las Nietas de Nonó.”
44 Apostolatos, “Foraging the Fragile and Fertile Soil.”
45 Irving, “Artists-in-residence Las Nietas de Nonó.”
46 Anzaldúa, “Making Choices,” 158.
47 Anzaldúa, “Making Choices,” 158.

48 Ofer and Alistar, "Felt Experiences with Kombucha Scoby," 15.
49 Barad, *Meeting the Universe Halfway,* 139.
50 da Silva, "On Difference without Separability," 59.

References

Anzaldúa, Gloria. *Borderlands/La Frontera: The New Mestiza*. 2nd ed. Aunt Lute Books, 1999.

Anzaldúa, Gloria. "Making Choices: Writing, Spirituality, Sexuality, and the Political: An Interview with AnaLouise Keating (1991)." In *Interviews/entrevistas: Gloria E. Anzaldúa*, edited by AnaLouise Keating. Routledge, 2000.

Apostolatos, Clara Maria. "Foraging the Fragile and Fertile Soil with Las Nietas de Nonó." *Intervenxions, the Latinx Project*, August 29, 2023. https://www.latinxproject.nyu.edu/intervenxions/foraging-the-fragile-and-fertile-soil-with-las-nietas-de-non.

Barad, Karen. *Meeting the Universe Halfway: Quantum Physics and the Entanglement of Matter and Meaning*. Duke University Press, 2007.

Boehmer, Elleke, and Dominic Davies, eds. *Planned Violence: Post/Colonial Urban Infrastructure, Literature and Culture*. Palgrave MacMillan, 2018.

Bureau of Land Management. "Infrastructure." Accessed June 19, 2024. https://web.archive.org/web/20240619214105/https://www.blm.gov/about/laws-and-regulations/infrastructure.

Casey, Edward S. "Walling Racialized Bodies Out Border versus Boundary at La Frontera." In *Living Alterities: Phenomenology, Embodiment, and Race*, edited by Emily S. Lee. State University of New York Press, 2014.

De León, Jason. *The Land of Open Graves: Living and Dying on the Migrant Trail.* University of California Press, 2015.

Deming, Alison H., and Lauret Savoy. "Widening the Frame." In *Colors of Nature*, edited by Alison Deming and Lauret Savoy. Milkweed, 2011.

Ferreira da Silva, Denise. "On Difference without Separability." *32nd Bienal de São Paulo Art Biennial: Incerteza Viva*. https://static1.squarespace.com/static/57be0e6bcd0f68f047b51971/t/6747cca1deb77d181d683e0b/1732758690541/D%2BFerreira%2Bda%2BSilva%2B %2BOn%2BDifference%2BWithout%2BSeparability.pdf.

Ford Foundation. "Las Nietas de Nonó." *Ford Foundation: The Art of Change*. Accessed October 2, 2017. https://www.fordfoundation.org/about/the-art-of-change-meet-our-fellows/las-nietas-de-nono/.

Huizar-Hernández, Anita. "Between Place and Plot: Reimagining the story of Arizona." *The Journal of Arizona History* 63, no. 3 (2022): 263–279. https://muse.jhu.edu/article/869735.

Irving, Jennifer. "Artists-In-Residence Las Nietas De Nonó Use Scobys to Represent Community in Visual Arts Center Exhibition." *College of Fine Arts, the University of Texas*, October 10, 2024. https://finearts.utexas.edu/news/artists-residence-las-nietas-de-nono-use-scobys-represent-community-visual-arts-center.

Koch, Natalie. *Arid Empire: The Entangled Fates of Arizona and Arabia*. Verso Books, 2023.

Lussier, Anna, and Peter Lindholm. "Rate of Aquatic and Maritime Drowning Deaths after US-Mexico Border Wall Height Increase." *JAMA: The Journal of the American Medical Association* 331, no. 12 (2024): 1059–1060. https://doi.org/10.1001/jama.2024.1275.

Mbembe, Achille. *Necropolitics*. Duke University Press, 2019.

Newns, Lucinda. "Necropolitical Ecologies: Creative Articulations of Nature's Death-Work in the Borderzone." *Interventions* 26, no. 1 (2023): 36–51. https://doi.org/10.1080/1369801X.2023.2190919.

Newns, Lucinda. "'The Sea Cannot Be Fenced': 'Natural' and 'Unnatural' Borders in Gloria Anzaldúa's *Borderlands/La Frontera* and Amitav Ghosh's *Gun Island*." *Interdisciplinary Studies in Literature and Environment* 29, no. 4 (2022): 1097–1120. https://doi.org/10.1093/isle/isab044.

Ofer, Netta, and Mirela Alistar. "Felt Experiences with Kombucha Scoby: Exploring First-Person Perspectives with Living Matter." In *Proceedings of the 2023 CHA Conference on Human Factors in Computing Systems*. ACM, 2023. https://doi.org/10.1145/3544548.3581276.

Office of the Texas Governor. "Operation Lone Star Boosts Border Response with New Marine Barriers." Accessed July 17, 2024. https://web.archive.org/web/20230717044838/https://gov.texas.gov/news/post/operation-lone-star-boosts-border-response-with-new-marine-barriers.

Rubenstein, Michael, Bruce Robbins, and Sophia Beal. "Infrastructuralism: An Introduction." *Modern Fiction Studies* 61, no. 4, 2015: 575–86. *JSTOR*. https://www.jstor.org/stable/26421821.

Saldaña-Portillo, María Josefina. "In the Shadow of NAFTA: 'Y Tu Mamá También' Revisits the National Allegory of Mexican Sovereignty." *American Quarterly* 57, no. 3 (2005): 751–777. https://doi.org/10.1353/aq.2005.0051.

Savoy, Lauret E. *Trace: Memory, History, Race, and the American Landscape*. Counterpoint, 2013.

Tamez, Margo. "Indigenous Women's Rivered Refusals in El Calaboz." *Diálogo* 19, no. 1 (2016): 7–21. https://doi.org/10.1353/dlg.2016.0030.

U.S. Customs and Border Protection. "Environmental Management." Accessed June 19, 2024. https://web.archive.org/web/20240619223744/https://www.cbp.gov/about/ environmental-management.

U.S. Customs and Border Protection. "U.S. Border Patrol Technology." Accessed June 19, 2024. https://web.archive.org/web/20240619011221/https://www.cbp.gov/border-security/along-us-borders/us-border-patrol-technology.

Zepeda, Ofelia. "Autobiography." In *Here First: Autobiographical Essays by Native American Writers*, edited by Arnold Krupat and Brian Swann. Modern Library, 2000.

Zepeda, Ofelia. "Birth Witness." In *Where Clouds Are Formed*. University of Arizona Press, 2008.

15 Soil Futures

Environmental Management as Care in Mary Ann Shadd Cary's "A Plea for Emigration"

Hannah Muhlfelder

Mary Ann Shadd Cary is known for her many intellectual legacies, though hardly for her contributions to environmental literature. Shadd Cary commented on a wide span of topics in her journalistic career, from emigration to Black radicalism and the role of women. Her complex persona looms large in the history of Black women's intellectualism in the nineteenth century.[1] Shadd Cary was the first Black woman to enroll in law school in the United States. She founded the *Provincial Freeman* newspaper and held intellectual sparring matches with the likes of Frederick Douglass and Martin Delany. She vigorously defended her convictions without mincing words. Shadd Cary was ahead of her time in many ways, including how she thought about land. Shadd Cary's writing about Black emigration to Canada demonstrates her environmental rationality, an understanding of natural resources grounded in careful observation.

Nineteenth-century racial oppression shaped Shadd Cary's views of land. As a free Black American, she understood that the Fugitive Slave Law of 1850 disintegrated her legal protections. In her 1851 pamphlet, "A Plea for Emigration, or Notes of Canada West," Shadd Cary encourages Black residents in the United States to emigrate to Canada, where slavery was illegal. Canada West, the region north of the Great Lakes, is now part of Ontario (see Figure 15.1). In 1852, Shadd Cary emigrated there to establish a school for fugitive slaves and their children. Shadd Cary believed Black settlement in Canada would bring racial uplift through public education and economic prosperity. She wrote, "I firmly believe that with an axe and a little energy, an independent position would result."[2] Cary believed the soil could help care for Black people escaping the Fugitive Slave Law. This chapter closely examines "A Plea for Emigration," exploring how Shadd Cary's radical abolitionist views shaped her ideas about land management and the natural environment. This analysis converges with recent works exploring Shadd Cary's emigrationism as Black feminist ecological thought, while adding further nuance to interpretations of Shadd Cary's text and particularly her emphasis on independence.[3]

DOI: 10.4324/9781003581185-19

Figure 15.1 George Woolworth, British, Danish, and Russian Possessions in North America, 34 centimeters by 42 centimeters, J.H. Colton Company, 1855. Here, Canada West is demarcated in darker gray.

Scholars have rightfully questioned the legitimacy of Shadd Cary's claims in "A Plea for Emigration." Some have felt that her depictions of fertile Canadian soil and mild climates were biased, because her pamphlet underrepresented the hardships that awaited Black emigrants when they arrived in Canada, like virulent racism and sparse resources.[4] However, I demonstrate that Shadd Cary's work possesses an environmental rationality beyond her idealism. Above all, Shadd Cary prioritizes what Nneka Dennie describes as Shadd Cary's "Black radical ethic of care."[5] While Shadd Cary's vision of Black settlement in Canada did not come into fruition as she planned, her arguments were based on accurate conclusions about the natural environment. Shadd Cary conducted extensive research on agriculture, forests, and soil health using "matter obtained both from individuals and from documents . . . of unquestionable character."[6] Her pamphlet illustrates Canadian ecosystems[7] as an important factor for survival, alongside networks of care. Viewing Shadd Cary in this way shows how she disrupts the largely white, masculine canon of nineteenth-century environmental literature and, instead, demonstrates study of the natural environment as part of a Black radical feminist intellectual tradition.

Shadd Cary chose to emigrate because she believed that survival was a mode of Black resistance against white institutions of violence. Nneka Dennie situates "a Black radical ethic of care" in Shadd Cary's work in a Black radical feminist intellectual tradition, alongside figures like Maria W. Stewart, Sarah Parker Remond, and Anna Julia Cooper. The "core" of this ethic of care, Dennie writes, involves "recognizing the multilayered forms of oppression that target Black communities; critiquing . . . antiblackness and white supremacy, and articulating or acting on a desire to realize change on behalf of Black communities."[8] Dennie frames Shadd Cary's work as an intellectual commitment to care. This analysis of "A Plea for Emigration" offers a case study for Dennie's framework. I illustrate how Shadd Cary expressed her "desire to realize change" not only in her rhetoric of moral uplift but also through her visualization of an agrarian future in which Black migrants sustained themselves. By imagining a brighter future in Canada, Shadd Cary engaged in a Black feminist practice of care. Here, as in many other points in history, Black women's care work under dire circumstances represents a refusal to submit to racial capitalism. Such care has fueled both intellectualism and practical work to sustain Black communities.[9] Shadd Cary's ethic of care was radical resistance, first through survival and second through reimagining and rebuilding a world beyond white supremacy.

While Shadd Cary's pamphlet shares generic similarities with settlement journals, as she encourages education and pays attention to agriculture and forestry, Shadd Cary's Black radical ethic of care causes her to diverge from settlement journals. In his 1998 republication of "A Plea for Emigration," Richard Almonte argues that Shadd Cary's encouragement of Black assimilation in Canada, though ethically problematic, is "a strictly utilitarian decision."[10] She believed that Black assimilation with white institutions would engender voting rights and paid employment.[11] Governments published settlement journals to encourage immigration, part of a settler-colonial campaign. Shadd Cary did not acknowledge the settler-colonial logics undergirding her view of Black settlement in Canada. She was, nevertheless, engaged in radical questions of how to make liberation happen, which led her toward rigorous research.[12] Shadd Cary, a lifelong academic, "endeavored to furnish information . . . from reliable sources and from observation."[13] However, Shadd Cary's conclusions concur with that of Canadian farmers who were writing at that time. Shadd Cary's description of the landscape in Canada may appear romanticized, but agricultural society documents echo Shadd Cary's favorable assessment of climate, soil, and agricultural outputs.[14] Shadd Cary dedicated herself to sharing accurate information, an act of care for people in dangerous circumstances.

This unique piece of nineteenth-century print culture disrupts canonical works of environmental thought from the period by illustrating land management as a survival tool for African American refugees. Shadd Cary's rhetoric deviates from the likes of George Perkins Marsh and Henry David Thoreau, who viewed rapid deforestation and urbanization as flagrant disrespect to nature.[15] Unlike white preservationists from the time, Shadd Cary advocated for active land management through forestry and agriculture. Although contemporary readers may associate her with political expertise, Shadd Cary was also highly knowledgeable about the natural world. Shadd Cary taught a botany class when she opened a school in Canada, making environmental education available to self-emancipated people and their children.[16] Nineteenth-century environmentalists like Thoreau believed that wilderness could provide solace for wealthy people escaping the ills of urban life. For Shadd Cary, the Canadian forest offered an escape from active threats against her life, and managing its lands would allow migrants to build stable communities.

Shadd Cary intimately understood the plight of emigrants. She moved several times over the course of her life, though her relocation to Canada shaped her political work most prominently. Born in 1823 to freedpeople in Wilmington, Delaware, Shadd Cary grew up with certain privileges. As Jane Rhodes explains in her biography of Shadd Cary's life, she "escaped the worst of slavery while living in a slave state, and benefited from a color-conscious social system."[17] However, strict racial codes threatened the Shadd family despite their status. The family moved to West Chester, Pennsylvania, in 1833, likely to avoid a law that barred Black students from attending public schools.[18] Shadd Cary moved to New York in 1851 to teach a primary school at a Black church. Later, to evade the Fugitive Slave Law, she emigrated to Canada, where she founded the *Provincial Freeman* newspaper. She used men's names as pseudonyms in hopes it would increase the paper's proliferation and "build community among Black Canadians."[19] Shadd Cary remained steadfast in her commitment to education and public intellectualism throughout her life. She charted new territory for Black women in the nineteenth century through her activism and willingness to transgress cultural, social, and political boundaries.

Shadd Cary's "A Plea for Emigration" earned her a place among high-profile abolitionists when it was published in 1852, and it remains one of her best-known works.[20] Shadd Cary spends the first 12 of the 44 pages of the pamphlet describing the landscape and natural resources of Canada West. She does this to illustrate how Black emigrants could be self-reliant through agriculture and land management. The first part of the pamphlet discusses Canada's climate, forestry, agriculture, and animal husbandry as an enticement to prospective emigrants. She then moves on to Ontario's trades, churches and schools, and politics and economy. Shadd Cary includes a section on Canadian laws of inheritance to demonstrate how Black emigrants may pass wealth to their children. She characterizes its immigrant population, which she says "consists of English, Scotch, French, Irish, and Americans."[21] Shadd Cary's pamphlet was structured to foreground land and environment, a crucial aspect of her writing that merits further study.

Shadd Cary closes by discussing other commonly suggested sites of Black emigration, like the British West Indies, Mexico, South America, and Africa, as well as Vancouver's Island in Canada. She encourages emigrants to "not be fastidious on emigrating to a country" if they "love liberty better than slavery."[22] While Shadd Cary points out the merits of other continents, she says "there seems to be no safe alternative left but to be satisfied with that government now existing that is most reliable and powerful."[23] Canada at the time was a British colony (see Figure 15.2). Shadd Cary viewed Canada as the most practical choice because the British government had passed the Slavery Abolition Act in 1833, although formerly enslaved people remained in unpaid apprenticeships until 1838.[24] As Charmaine Nelson points out, Canada remained a space of slavery for many fugitives seeking safety there.[25] Shadd Cary viewed the paltry legal freedoms available in Canada as a safer possibility for African Americans escaping the Fugitive Slave Law in the United States. With proslavery Democrats in power in the US government, Shadd Cary thought that legal protections in Canada were more stable.

Figure 15.2 John Tallis, Map of West Canada or Ontario, c. 1850, 10 inches by 13 inches, in *The Illustrated Atlas, And Modern History Of The World Geographical, Political, Commercial & Statistical*, edited By R. Montgomery Martin, Esq., 1851.[26]

Shadd Cary understood that emigration posed considerable challenges but felt that Canada offered the best shield once the United States passed the Fugitive Slave Law of 1850. Shadd Cary acknowledges rampant prejudice toward Black people in Canada, particularly among white French and "Yankee" immigrant populations.[27] Nevertheless, she concludes that "no settled country in America offers stronger inducements to colored people."[28] She uses the word "America" to describe the whole of the North American continent, unseating the idea of US continental dominance. Shadd Cary noted in her pamphlet that racial violence had continued to escalate in the United States in the recent years and felt remaining stateside was no longer worth the risk.[29] She viewed Canada as protected by "the strong arm of British power,"[30] though she never discussed the British empire's violent history of colonialism. Shadd Cary's treatise replicated the logics of settler colonialism, though her critiques, as R.J. Boutelle and Marlas Yvonne Whitney write, "provided a starting point from which subsequent generations have envisioned more liberatory possibilities."[31] Shadd Cary did not idealize Canada as an oasis for Black emigrants, but instead viewed the country as the best option among the United States's closest neighbors for survival.

Shadd Cary promotes emigration to Western Ontario over other parts of Canada like Vancouver Island. She discusses Vancouver Island's climate, which she compares to the Mediterranean. Shadd Cary writes that the soil there "is of the best description" and that its location along trade routes offers economic advantages.[32] Despite the merits of Vancouver Island, Shadd Cary was more focused on Canada West.[33] Black people in the United States could travel to cities in Canada West like Toronto more easily than to Vancouver Island, which lay on the other side of the continent. When white residents of Cincinnati rioted in 1829, they violently expelled their Black neighbors. African Americans escaped to Canada West, establishing a colony they named Wilberforce, after the Canadian abolitionist William Wilberforce.[34] Abolitionists formed networks along the route to enclaves like Wilberforce along the Underground Railroad. Toronto was a well-known terminus for enslaved people attempting to self-emancipate.[35] Shadd Cary's focus on Canada West suggests that she published her pamphlet to help enslaved people and freed people alike escape the Fugitive Slave Law. Shadd Cary's interest in Canada West was influenced by geography, revealing the breadth of her target audience.

The author addresses climate to subvert the idea that Black people could only thrive in southern climes. Shadd Cary assures her readers that the climate in Canada West is milder than Americans may assume. She writes, "I have thought proper to allude to the cold, at first, for it is the feature in the climate most dwelt upon," particularly by friends and members of her community who she says are "ignorant on this point."[36] Shadd Cary implies that people have refused to emigrate to Canada based on temperature alone. Of Canada West, Shadd Cary says that it is "exempt from the steady and enfeebling warmth of southern latitudes, and the equally injurious characteristics of polar countries," which she deems "highly conducive to mental and physical energy."[37] Shadd Cary thought that either climactic extreme would be "injurious," but that the temperate climate of Canada West was livable. She depicts how emigrants would channel their "mental and physical energy" into their new lives. In this description, Shadd Cary rebuts popular and long-held racist belief that Black people were better suited to warm climates than their white counterparts.[38] The author declares Canada West as a climatically conservative choice, subverting dominant ideologies about Canadian climate and reasserting African Americans' personhood.

Shadd Cary's climatic assessment corresponds with reports from Canadian farmers during this period, the result of her careful research. An 1861 settlement pamphlet from the Canadian government states that the "climate of Canada is in some measure exceptional, especially that of the Peninsular portion. The influence of the great [sic] Lakes is very strikingly felt in the elevation of winter temperatures and in the reduction of summer heats."[39] Shadd Cary draws the same sketch about the Great Lakes with person-first language, stating that "people living in the states bordering [the Lakes], suffer more severely from the cold than Canadians."[40] Farmers in mid-nineteenth-century Ontario wrote that their region had "advantages of climate and soil" that other regions of Canada did not possess.[41] Shadd Cary's belief about Toronto's mild climate was not an idealized image based on a promise of racial equality, but rather an observation made from the information available to her. This climatic analysis exemplifies her awareness of the Canadian environment and her careful consideration toward emigrants.

Much of Shadd Cary's optimism about Black futures in Canada lies in the quality of the soil, which she views as a key indicator of overall ecosystem health.[42] She mentions soil quality several times in this section of the pamphlet.[43] Shadd Cary points out that soil health influences the viability of crops, the quality of dairy products due to the health of pastures, as well as the condition of trees and forests. She illustrates soil as part of an interconnected system of physical environment and biota, what would later be described as an ecosystem. The benefits of these relationships, or ecosystem services, could become sources of income in Shadd Cary's eyes. Shadd Cary observed the soils in Canada during her travels, noticing its "dark and heavy appearance, and the depth."[44] Emphasizing the importance of high quality soil, Shadd Cary maintains that Black people's "welfare, in a pecuniary view, is promoted by attention to the quality of the soil."[45] The economic benefits of soil came from harvesting crops or timber. Shadd Cary's emphasis on soil demonstrates her understanding of land management as the product of dependent relationships within an ecosystem.

Shadd Cary highlights the importance of caring for the soil through farming, offering suggestions to Black emigrants managing their land. She felt that emigrants should not take soil quality for granted but rather continue to care for it. Such reciprocal relationships between farmers and soil have long existed.[46] One farmer in the 1848 *Agricultural Journal of the Lower Canada*

Agricultural Society explained that manures could be used to care for the soil, but that they contained a "large quantity of ammonia, carbonic acid, &c."[47] The writer cautioned other growers to use these fertilizers in moderation to prevent chemical imbalances in soil. Farmers in Canada had an intimate knowledge of their soils and shared information with one another to improve yields. Shadd Cary's promotion of "attention" to soil calls upon Black emigrants to manage land carefully to ensure high yields, promoting economic stability. Shadd Cary viewed land management as a reciprocal network of care including people and land. Her dedication to Black futures separates her from many environmental writers, who tended to view wilderness as separate from human societies.

Shadd Cary encouraged emigrants to manage their forests and agricultural lands to ensure a quicker return on their investments. Timber, for example, offered substantial cash flow: "more may be realized from firewood than to three times pay the cost of a farm."[48] She lists the kinds of trees growing in Ontario's forests, including "walnut, hickory, white and burr oak, basswood, ash, pine, and poplar—all of the largest size."[49] Her description indicates that she understood how to identify different trees or did the research necessary to understand the species makeup of the forests. Shadd Cary points out that timber offers a faster return than agricultural products, which take time to grow and harvest. She juxtaposes this passage with a description of the tax system in Canada, with a special focus on family inheritance. Shadd Cary wants Black emigrants in Canada to build generational wealth and to do so through independently owned farms and timber operations. She imagined people managing land over the long term, securing their families' futures. Shadd Cary writes about forestry as an opportunity for emigrants to build wealth that would not have been possible under the US proslavery government.

Despite Shadd Cary's focus on land management as a kind of business venture, she also viewed agriculture as an important intellectual and cultural way of life. She writes of Canadian farmers' self-assessment mechanisms, "there are, and must be for a time, few experimental and scientific farmers, as it is more a means of present subsistence." However, she felt that "the conviction is irresistible that indigence and moderate competence must at no distant day, give place to wealth, intelligence, and their concomitants."[50] Shadd Cary describes the stages of settlement in Ontario not as urbanization, where industry would replace subsistence farming, but instead as an increasingly scientific rural community. Such "experimental and scientific" farming might include testing different seed varieties to improve crop yields or developing new varieties of fruit.[51] In the mid-nineteenth century, the advent of railroads in Toronto began to transform mass transit, trade, and exports.[52] Shadd Cary hoped that Black emigrants would soon benefit from these changes and be able to export their agricultural products. Although her work suggests that the only worthy future for Black emigrants was one driven by capital and scientific improvement, racial uplift in Shadd Cary's work represents the desire to create a better, more equal world.

Shadd Cary valued not only intellectual pursuits in the life of a Canadian farmer but also its resulting diet. She declares that the quality and size of the vegetables at the markets she visited in Canada West were the likes of which she had "never seen in the large markets of our northern cities."[53] Shadd Cary lists the fruits and vegetables available, elaborating far beyond the staple crops mentioned in other settlement journals:

> [T]omatoes, cucumbers, onions, beets, cabbage and cauliflower, egg-plants, beans, peas, leeks, celery, lettuce, asparagus, melons, water-melons and musk-melons, canteloupes [sic] and spinach. There are other vegetables, but they have been mentioned elsewhere.[54]

She deftly determines the kinds of produce that would be available all through the year, demonstrating how emigrants in Canada West would have a seasonal and varied diet. Shadd Cary's list includes foods that would be familiar to US Southerners, who might have been surprised to find Canadian soil less barren than they imagined. Saying only that "they have been mentioned elsewhere," the author indicates that not only has she done research, but that her reader should also read about Canada West in preparation to emigrate. In this vignette of Canadian life, Shadd Cary anticipates healthy soils providing varied and nourishing food that would be familiar to US emigrants.

Shadd Cary's politics influenced her descriptions of agriculture, though her politics cause her to diverge from such publications. Settlement journals sometimes mentioned little more than the maximum yield per acre of wheat. Some mention the popularity of grain vegetables and the cost of seeds to grow them.[55] Offering more detail, Shadd Cary uses a table to describe average yield (see Figure 15.3).

GRAINS, POTATOES, TURNIPS, &C.

The accompanying table exhibits the average yield to the acre, of the several grains mentioned, in fallow land:

ARTICLES.	NO. BUSH.	ARTICLES.	NO. BUSH.
Wheat,	30	Oats,	70
Buckwheat,	15	Barley,	40
Rye,	35	Indian Corn,	50

Figure 15.3 Mary Ann Shadd Cary, "Table of Average Yield to the Acre of Grains, Potatoes, Turnips, &c.," "A Plea for Emigration," page 11, 1851.

Yield by acre helps prospective emigrants make useful calculations about their estimated profits from farming after investing funds in seeds, land, and farming equipment.[56] A thoughtful use of data would help families create a budget to prevent food shortages, yet another act of care. If Shadd Cary was merely focused on persuading people of the possibilities of Canadian life, she could have used the maximum yield per acre. Shadd Cary's priorities in information sharing reflect a Black radical ethic of care as she aimed to help families prevent food shortages on their new farms.

Shadd Cary shared her research observations in hopes emigrants would become self-reliant. The author stresses the value of independence throughout the pamphlet, which Almonte has analyzed as having two functions: "to do things by one's self, but more importantly, independently of white people."[57] However, Shadd Cary encouraged people to live in integrated communities. In Shadd Cary's case, independence was about avoiding intervention from white institutions, looking instead to the soil as a more efficient and reliable source of economic aid. Shadd Cary's insistence on Black people integrating with white people is a form of racial uplift; her writing echoes her belief that Black people needed to assimilate to thrive. Though this view was erroneous, her emphasis on self-reliance represented an attempt to shield emigrants from untrustworthy institutions. Rather, the land deserved one's trust. The ecology of Canada as envisioned by Shadd Cary stipulates self-reliance while forming life-sustaining management relationships with land.

Shadd Cary's writing about land purchase offers further nuance to her view of economic independence. Despite Shadd Cary's discouragement of using government funds, she encouraged emigrants to purchase land from the government instead of from individuals. She insisted that the 100-acre lots parceled by the Canadian colonial government sold at better rates than the individual plots for sale that were significantly smaller, but priced higher per acre. Shadd Cary writes that "too many are now *independently* dragging along miserably" and that significant acreage is required to have a "comfortable support" (emphasis original).[58] The use of italics in this phrase offers an ironic twist on the use of independence elsewhere in the pamphlet, suggesting that independence means little without a comfortable lifestyle. Although Shadd Cary discouraged settlements based on Black self-determination like Wilberforce, she believed that it was better to thrive in interdependence than suffer in independence. This approach to land represents a more balanced view of independence, one less clouded by the ideological problems of racial uplift.

Shadd Cary's fierce advocacy of flight forcefully rejected proslavery legislation in favor of Black resistance. As historian Kellie Carter Jackson writes in *Force and Freedom: Black Abolitionists and the Politics of Violence,* "with emigration, force was flight. The withdrawal of black labor and black bodies was an attack on the slaveholding class. In effect, to 'steal away' was a direct personal and political tool."[59] Carter Jackson articulates force as a method of transformative action. In an economic system where slaveholders exploited

Black people for enormous profit, enslaved people escaping to freedom forcefully removed wealth from the "slaveholding class." The ability to escape or emigrate was a "personal and political tool," the power of which Shadd Cary publicly wielded. Providing information as to how someone would subsist after emigrating offered support to help people flee or prevent enslavement. Shadd Cary's advocacy of emigration as flight and her research about how environmental resources could mitigate risk represent a choice we can locate centuries later as part of a Black radical tradition.

African American environmental thinkers like Shadd Cary have long articulated soil as a tool to heal from colonial and racial violence. Fannie Lou Hamer's work on the Freedom Farms Cooperative, established in 1967, offered a means for African Americans to exit the exploitative sharecropping system while staying in the South and building healthy communities.[60] Dianne Glave and Mark Stoll, in their "historiography of African American environmental history," explain how Black people, and particularly Black women, used environmental knowledge to advocate for their families and communities.[61] Tiya Miles, in her biography of Harriet Tubman, calls her "the most famous Black woman ecologist in U.S. history," someone who was "capable of picturing possible futures that contradicted present realities."[62] Leah Penniman explains that soil requires deep care and "the restoring of organic matter to the soil was part of healing from colonialism."[63] As Eunice Toh writes about Shadd Cary's resonance in contemporary Detroit, "it was not a returning or a restoring, but a *recovering* of land that persists with the Black urban farmers of Detroit."[64] African American women activists have and continue to articulate Black agrarianism as a means of establishing self-sovereignty.

Shadd Cary's work and the Black radical feminist tradition of environmental thought to which she belongs intervene on canonical works of environmental literature. While nineteenth-century writers like George Perkins Marsh believed that humans ought to be cautioned for cutting down trees or using large swaths of land for agriculture,[65] Shadd Cary felt that land could be managed and developed efficiently. Henry David Thoreau's escape from "the improvements of ages" in *Walden; or, Life in the Woods* contrasts strongly with Shadd Cary's encouragement of settlement through farming.[66] Her work looks up to land management through sustainable agriculture and forestry as flight, akin to Thoreau's escape to Walden. However, unlike her white-male counterparts, Shadd Cary did not leave New York City to transcend the ills of urban life, but rather to survive changes in legislation that could prove lethal to African Americans. Though entangled in systems of settler colonialism, Shadd Cary's view of environmental resource management was predicated on resistance and a radical ethic of care.

"A Plea for Emigration" diverges from environmental literature of the period that prioritizes natural resource preservation, considering land management in a different way. In his keyword entry to this volume, Joshua T. Anderson writes that "management reminds us the power to change our story is in our hands." Shadd Cary believed the same, advocating fiercely for Black emigration and agrarianism in Canada as a means of establishing new environmental futures. She illustrates Black people as environmental managers, an important expression of power in a place and time in which land dictated one's personal and economic autonomy. Shadd Cary's pursuit of information sharing exhibits how scholarship about land management can precipitate radical change. Shadd Cary's insistence on emigration haunts us when communities of color are made to absorb the worst impacts of climate change. Today, her work offers a call to action to care for the planet and one another in the face of disaster. "A Plea for Emigration" illustrates how looking away from the canon may give way to new modes of thinking. Highlighting survival, stability, and recovery in Shadd Cary's work suggests that an important practical and theoretical tool to adapt to climate futures may lie in the writing of the past.

Notes

1 Some other works on Mary Ann Shadd Cary's impact on Black radicalism include *Mary Ann Shadd Cary*; Streitmatter, "Mary Ann Shadd Cary," 25–36; Peterson, *Doers of the Word*; Conaway, "Rhetorically Constructed Africana Mothering in the Antebellum"; Rhodes, *Mary Ann Shadd Cary*; and Moriah, ed., *Insensible of Boundaries*.
2 Shadd Cary, "A Plea for Emigration."
3 For more on Shadd Cary's ecological thought, see Toh, "Mary Ann Shadd Cary's Black Soil Ecology," 77–94; Boutelle and Whitley, "Plotting New Gardens," 145–164.
4 See Rhodes, *Mary Ann Shadd Cary*; Bain, "Black Emigrant in Canada," 183–186.
5 Shadd Cary, *Mary Ann Shadd Cary*.
6 Shadd Cary, "A Plea for Emigration," iv.
7 In 1935, Arthur Tansley first described an ecosystem as a place-specific relationship between living things and their physical environment. As Shadd Cary describes this relationship without using the term, I have chosen to use the word "ecosystem" to situate her in contemporary scientific inquiry. For more on ecosystems and an emphasis on nonhuman life, see Johnson, "Abiotic Ecosystems?," 39–53.
8 Shadd Cary, *Mary Ann Shadd Cary*, 11.
9 Free Black people in the North in the mid-nineteenth century relied on networks of care to survive when means were limited. For more, see Horton and Horton, *In Hope of Liberty*; Schweninger, "Property Owning Free African-American Women in the South, 1800–1870"; Glymph, *Out of the House of Bondage*. Glymph's study of the "work and thriftiness of Black women" during the Reconstruction Era offers an important framework for Black radical feminism in Shadd Cary's scholarship.
10 Shadd Cary, "A Plea for Emigration."
11 Shadd Cary, "A Plea for Emigration," 10.
12 For more on Shadd Cary's intended audience and the use of settlement journals as genre, see "A Plea for Emigration; or, Notes of Canada West."
13 Shadd Cary, "A Plea for Emigration," iv.
14 *Agricultural Journal and Transactions of the Lower Canada Agricultural Society*, 1–381.
15 Marsh, *Man and Nature*; Thoreau, *Walden*, 1854.
16 Rhodes, *Mary Ann Shadd Cary*.
17 Rhodes, *Mary Ann Shadd Cary*, 5.

18 See Streitmatter, "Mary Ann Shadd Cary: Advocate for Canadian Emigration"; Shadd Cary, *Mary Ann Shadd Cary*, 3.
19 Shadd Cary, *Mary Ann Shadd Cary*, 4.
20 Shadd Cary, *Mary Ann Shadd Cary*.
21 Shadd Cary, "A Plea for Emigration," 34.
22 Shadd Cary, "A Plea for Emigration," 39.
23 Shadd Cary, "A Plea for Emigration," 40.
24 "The 1833 Abolition of Slavery Act and Compensation Claims."
25 Charmaine Nelson, "The Canadian Fugitive Slave Archive and the Concept of Refuge," *English Studies in Canada* 45, no. 3 (2019): 91–115.
26 John Tallis' beautifully rendered but paradoxical map depicts Canada West or Ontario as an organized colony and a wild, sublime landscape. This map both romanticizes First Nations peoples and eliminates Native place-making, visually marginalizing "Indian Territory" beyond the map's edge.
27 Shadd Cary, "A Plea for Emigration," 33–35. Here, "Yankee" most likely refers to white American immigrants in Canada.
28 Ibid, 36.
29 Ibid, 44.
30 Ibid.
31 Boutelle and Whitley, "Plotting New Gardens," 161.
32 Shadd Cary, "A Plea for Emigration," 43.
33 Shadd Cary, "A Plea for Emigration," 36.
34 Taylor, "Reconsidering the 'Forced' Exodus of 1829," 283–302.
35 Pollock-Ellwand, "Travelling the Route from Designation to Local Action," 372–388.
36 Shadd Cary, "A Plea for Emigration," 6.
37 Shadd Cary, "A Plea for Emigration," 6.
38 Asaka, *Tropical Freedom*.
39 Lovell, *Canada*.
40 "A Plea for Emigration," 6. This comparison between the American and Canadian sides of the Great Lakes appears pervasive among Canadian settlement journals, indicating that Shadd Cary was engaged in the information sharing of her moment.
41 *Agricultural Journal and Transactions of the Lower Canada Agricultural Society*, 1.
42 For more on the use of Black soil ecologies in "A Plea for Emigration," see Toh, "Mary Ann Shadd Cary's Black Soil Ecology."
43 Shadd Cary, "A Plea for Emigration."
44 Shadd Cary, "A Plea for Emigration."
45 Shadd Cary, "A Plea for Emigration," 8.
46 Güldner, Larsen, and Cunfer, "Soil Fertility Transitions in the Context of Industrialization, 1750–2000," 785–811.
47 *Agricultural Journal and Transactions of the Lower Canada Agricultural Society*, 7.
48 Shadd Cary, "A Plea for Emigration," 10.
49 Shadd Cary, "A Plea for Emigration," 9
50 Shadd Cary, "A Plea for Emigration," 11.
51 Shadd Cary, "A Plea for Emigration," 11.
52 For more on the nineteenth-century history of Canada West, see Canniff, *History of the Province of Ontario (Upper Canada) Containing a Sketch of Franco-Canadian History—the Bloody Battles of the French and Indians—the American Revolution—the Settlement of the Country by U.E. Loyalists—Trials and Hard-Ships Incident to Pioneer Life—Thrilling Narratives of Adventures with Indians and Wild Beasts—Struggles for the Establishment of Christianity and Schools—Agriculture and Manufactures—the War of 1812—Battle Scenes, Capture of Buffalo et Al*; Goheen, "Victorian Toronto, 1850–1900."
53 Shadd Cary, "A Plea for Emigration," 11.

54 Shadd Cary, "A Plea for Emigration," 12.
55 For examples of discussions of settlement and maximum crop yield, see Lovell, *Canada*; Brown, *Views of Canada and the Colonists Embracing the Experience of a Residence, Views of the Present State Progress, and Prospects of the Colony, with Detailed and Practical Information for Intending Emigrants*; and Canniff, *History of the Province of Ontario (Upper Canada) Containing a Sketch of Franco-Canadian History et al.*
56 Shadd Cary, "A Plea for Emigration," 11.
57 Almonte, "A Scholarly Edition of Mary A. Shadd Cary's a Plea for Emigration, or Notes of Canada West."
58 Shadd Cary, "A Plea for Emigration," 18.
59 Jackson, *Force and Freedom.*
60 White, *Freedom Farmers.*
61 Glave and Stoll, *To Love the Wind and the Rain*, 5–8.
62 Miles, *Night Flyer.*
63 Penniman, *Farming While Black*, 123.
64 Toh, "Mary Ann Shadd Cary's Black Soil Ecology," 88.
65 Marsh, *Man and Nature.*
66 Thoreau, *Walden; or, Life in the Woods.*

Bibliography

Agricultural Journal and Transactions of the Lower Canada Agricultural Society 1, no. 1 (January 1848): 1–381. https://babel.hathitrust.org/cgi/pt?id=coo.31924086750258&seq=28

Almonte, Richard. "A Scholarly Edition of Mary A. Shadd Cary's A Plea for Emigration, or Notes of Canada West." PhD Diss., Concordia University, 1996.

Asaka, Ikuko. *Tropical Freedom: Climate, Settler Colonialism, and Black Exclusion in the Age of Emancipation.* Duke University Press, 2017.

Bain, Tracy. "Black Emigrant in Canada." *Canadian Literature*, no. 173 (2002): 183–186.

Boutelle, R. J., and Marlas Yvonne Whitley. "Plotting New Gardens: The Black Feminist Roots of Community-Building in Plea for Emigration." In *Insensible of Boundaries: Studies in Mary Ann Shadd Cary*, edited by Kristen Moriah, 145–164. University of Pennsylvania Press, 2025.

Brown, J. B. *Views of Canada and the Colonists Embracing the Experience of a Residence, Views of the Present State Progress, and Prospects of the Colony, with Detailed and Practical Information for Intending Emigrants.* Adam and Charles Black, 1844.

Conaway, Carol B. "Rhetorically Constructed Africana Mothering in the Antebellum: The Racial Uplift Tradition of Mary Ann Shadd Cary." *Journal of Pan African Studies* 2 (2007): 4–18.

Glave, Dianne D., and Mark Stoll, eds. *To Love the Wind and the Rain: African Americans and Environmental History.* 1st ed. University of Pittsburgh Press, 2005.

Goheen, Peter G. "Victorian Toronto, 1850–1900: Pattern and Process of Growth." PhD diss., University of Chicago, 1970. ProQuest (T-24319).

Güldner, Dino, Laura Larsen, and Geoff Cunfer. "Soil Fertility Transitions in the Context of Industrialization, 1750–2000." *Social Science History* 45, no. 4 (August 20, 2021): 785–811. https://doi.org/10.1017/ssh.2021.26.

Horton, James Oliver, and Lois E. Horton. *In Hope of Liberty: Culture, Community, and Protest among Northern Free Blacks, 1700–1860.* Oxford University Press, 1998.

Jackson, Kellie Carter. *Force and Freedom: Black Abolitionists and the Politics of Violence.* 1st ed. University of Pennsylvania Press, 2019.

Johnson, Benn. "Abiotic Ecosystems? A Critical Examination of Arthur Tansley's Ecosystem Definition." *Environmental Ethics* 42, no. 1 (January 1, 2020): 39–53. https://doi.org/10.5840/enviroethics20204215.

Lovell, John. *Canada: A Brief Outline of Her Geographical Position, Productions, Climate, Capabilities, Educational and Municipal Institutions, Fisheries, Railroads, &c, &c, &c.* Quebec: Authority, 1861.

Marsh, George Perkins. *Man and Nature; or, Physical Geography as Modified by Human Action.* Weyerhaeuser Environmental Classics. Charles Scribner & Co, 1867.

Miles, Tiya. *Night Flyer: Harriet Tubman and the Faith Dreams of a Free People: Harriet Tubman and the Faith Dreams of a Free People.* Significations. Penguin Press, 2024.

Moriah, Kristen, ed. *Insensible of Boundaries: Studies in Mary Ann Shadd Cary.* University of Pennsylvania Press, 2025.

Penniman, Leah. *Farming While Black: Soul Fire Farm's Practical Guide to Liberation on the Land.* Chelsea Green Publishing, 2018.

Peterson, Carla L. *Doers of the Word: African-American Women Speakers and Writers in the North (1830–1880).* Race and American Culture. Oxford University Press, 1995.

Pollock-Ellwand, Nancy. "Travelling the Route from Designation to Local Action: The Case of the Underground Railroad Settlement in Buxton, Ontario, Canada." *International Journal of Heritage Studies* 12, no. 4 (July 1, 2006): 372–388. https://doi.org/10.1080/13527250600726937.

Rhodes, Jane. *Mary Ann Shadd Cary: The Black Press and Protest in the Nineteenth Century.* New ed. Indiana University Press, 2023.

Shadd Cary, Mary Ann. *Mary Ann Shadd Cary: Essential Writings of a Nineteenth-Century Black Radical Feminist*, edited by Nneka D. Dennie. Oxford University Press, 2023. https://doi.org/10.1093/oso/9780197609460.001.0001.

Shadd Cary, Mary Ann. *A Plea for Emigration: Or, Notes of Canada West.* 1st ed. George W. Pattison, 1852.

Shadd Cary, Mary Ann. *A Plea for Emigration: Or, Notes of Canada West*, edited by Richard Almonte. Mercury Press, 1998.

Shadd Cary, Mary Ann. *A Plea for Emigration: Or, Notes of Canada West*, edited by Phanuel Antwi. Broadview Press, 2016.

Streitmatter, Rodger. "Mary Ann Shadd Cary: Advocate for Canadian Emigration." In *Raising Her Voice: African-American Journalists Who Changed History.* 1st ed., 25–36. University Press of Kentucky, 1994.

Taylor, Nikki. "Reconsidering the 'Forced' Exodus of 1829: Free Black Emigration from Cincinnati, Ohio to Wilberforce, Canada." *The Journal of African American History* 87, no. 3 (July 1, 2002): 283–302. https://doi.org/10.2307/1562479.

Thoreau, Henry David. *Walden; or, Life in the Woods.* Ticknor and Fields, 1854.

Toh, Eunice. "Mary Ann Shadd Cary's Black Soil Ecology." In *Insensible of Boundaries: Studies in Mary Ann Shadd Cary*, edited by Kristen Moriah, 77–91. University of Pennsylvania Press, 2025.

White, Monica Marie. *Freedom Farmers: Agricultural Resistance and the Black Freedom Movement.* 1st ed. Justice, Power, and Politics. University of North Carolina Press, 2018.

William Canniff. *History of the Province of Ontario (Upper Canada) Containing a Sketch of Franco Canadian History—the Bloody Battles of the French and Indians—the American Revolution—the Settlement of the Country by U.E. Loyalists—Trials and Hard-Ships Incident to Pioneer Life—Thrilling Narratives of Adventures with Indians and Wild Beasts—Struggles for the Establishment of Christianity and Schools—Agriculture and Manufactures—the War of 1812—Battle Scenes, Capture of Buffalo et Al.* A.H. Hovey, 34 King Street West, 1872.

16 "Back to Belonging"

Sound, *Ceremony*, and Resonance in a Re-Membered Commons

Kristen Brown

For the first time, and very belatedly, the Intergovernmental Panel on Climate Change recognized that "historical and ongoing patterns" of colonialism have directly exacerbated the effects of climate change.[1] Historical data extending carbon accounting beyond fossil fuels to include land use and forestry—thereby more accurately capturing the environmental destruction of colonial deforestation, agricultural conversion, and extractive land management practices—"significantly alters the top 10" emitters, though the United States remains by far the largest historical contributor, with China a distant second.[2] In the North American Southwest, evidence abounds for the settler state's poorly calculated extraction-based management of lands and waters. Mines, old and new, scar the landscape, from so-called "energy sacrifice zones" on sacred sites to over 500 abandoned uranium mines in the Navajo Nation alone.[3]

Colonial-capitalist institutions normalize ideological frameworks of dominance that fracture our understanding of ecological interdependence. The ongoing legacies of these violent interactions with the planet deeply impact environmental and human health. Despite a dualist ontology rooted in Euro-Western thinking, the "environment" and the "human" are, we might remember, inextricably woven and reciprocal systems. We can take heart, however, in the fact that this separation is relatively recent. Many Indigenous worldviews have long recognized the sentience and agency of nonhuman kin in reciprocal webs of belonging, as Lisa Fink and Lydia Cagluaq Agnus beautifully illuminate in their keyword entry for *relationality*. Settler-colonial ideologies condition us to forget this sameness and subsequently render "the environment" a separate entity, "a source of belongings," out there, for use and accumulation. We forget, notes Robin Kimmerer, that these lands and waters can—and should be—"the source of our most profound sense of belonging."[4] This separation, what Tewa Pueblo scholar Greg Cajete refers to as "the post-modern crisis of environmental perception," neglects embodied practices and sensory ways of knowing that have sustained ecological communities for millennia.[5]

This ontological shift from "belongings" to "belonging" might be facilitated, in part, by what Pauline Oliveros calls Deep Listening, a practice that

DOI: 10.4324/9781003581185-20

provides "the way to connect with the acoustic environment, all that inhabits it, and all that there is . . . [it is] the key to multi-level existence . . .[and] includes all sounds expanding the boundaries of perception."[6] This practice, notes Ellen Waterman, "troubles the self/world distinction so central to settler ideologies."[7] Water protectors at Standing Rock remind us of this collective intimacy with the singing, drumming, and ceremonial enactment of Mní Wičóni. Water is life. These sonic protocols, created to nourish relationships with place, were met with sonic violence in the form of Long Range Acoustic Devices (LRAD) deployed to sever those connections and instead protect the movement of crude oil through a 1,172-mile pipeline that crosses waterways 202 times.[8] The stark contrast between ceremonial song and weaponized sound reveals how acoustic practices—and the quality of listening they invite or foreclose—can either cultivate the relational intimacy necessary for responsible ecological stewardship or actively dismantle the very connections upon which that responsibility depends.

This systemic denial of Indigenous water sovereignty reflects broader patterns of colonial resource extraction and legal marginalization in North American management paradigms. The initiation of the Colorado River Irrigation Project in 1867, followed by the 1883 Religious Crimes Code that criminalized Indigenous song and ceremony for nearly a century, imposed a scarcity mindset reinforced by attempts to sever multigenerational and deeply embodied Indigenous management of life-giving waters. Cajete explains that, for Indigenous peoples, "compacts must be made between sources of life" with "key relationships . . . reflected in ceremonies."[9] During this ban on ceremonial practices, a strikingly different compact emerged, the Colorado River Compact, initiated in 1922. Rather than relationships of care developed over many generations of observation and engagement, this state-sanctioned compact is a complex tangle of regulative resource management steeped in notions of ownership (senior rights, paper rights, wet water rights) that defer to "expeditious agricultural and industrial development," thereby imposing an axiological framework well aligned with settler-colonial civilization discourse and its capitalist priorities.[10] The deliberate suppression of Indigenous ceremony—from the 1883 ban to the LRAD at Standing Rock—violently disrupts sensory practices of reciprocal relations embodied in and activated by ceremony. Shawn Wilson (Opaskwayak Cree), whose *Research is Ceremony* sheds light on these relationships, asserts that ceremony "allows us a raised level of consciousness and insight into our world," thereby revealing the many registers of violence in ceremonial suppression, not the least of which is epistemic.[11] Dylan Robinson (Stó:lō) notes that the prohibition of ceremonial potlatches and winter dances in Canada "was essentially the equivalent of banning books that document law and history."[12] The ecological knowledge woven through ceremonies underscores the need for Indigenous-led stewardship that amplifies the voices of lands, waters, and cosmologies.

While Indigenous peoples have lived in, on, and with what is now known as the Colorado River Basin for millennia, developing deep knowledge and sustainable relations over generations, less than two centuries of settler-colonial mismanagement have led to dangerous water shortages. Conflicts are all-too-often left to courtroom interpretation, a fraught space where legal practitioners often lack awareness of Indian Law, much less the inclination to consider the river's own laws, woven, in part, through stories and ceremony. Decades of unresolved tribal water rights reveal the inequity impacting Indigenous Nations. A 2022 report notes that

> Navajo [Diné] residents are 67 times more likely than other Americans to live without access to running water . . . [and] approximately 75 percent of people living on Hopi land are drinking contaminated water, which poses serious public health risks to the community.[13]

In 2023, the Supreme Court ruled against the Navajo Nation in *Arizona v. Navajo Nation*, a long-delayed water rights dispute, thereby failing to uphold both the 1868 Treaty of Bosque Redondo promising to provide resources and the 1908 Winters Doctrine stating that reservations have a right to water and may have priority in times of shortage. As of this writing, an ongoing water rights settlement involving three Tribal Nations—the largest in US history—raises concerns of insufficient infrastructure and predatory leasing frameworks.[14] The fundamental distinction between water as life—sustained through ceremony, story, and relational stewardship—and water as object—commodified through litigation, appropriation, and bureaucratic control—signals the unsustainable inequity of Western management paradigms and the urgent need for Indigenous-led stewardship.

Within these proliferating environmental crises, Indigenous scholars and knowledge-keepers offer alternatives that challenge the foundational assumptions of settler state dominance and commodification. Cajete diagnoses the ontological dimensions of these crises, noting that "our current worldview causes escalation of our destructive relationship with the environment at every level of life" and proposes a "modern ecosophy [that] would be about the rediscovery of meaning as it relates to our universe. It would require not only a different way of thinking, but also a different way of knowing and living."[15] This engagement with or, more accurately, this engagement *as* the natural world has already been here; however, for many, it requires a "rediscovery" or remembering. Cajete also highlights the difference between *thinking* and *knowing*, an important distinction elided by the conditioning of Cartesian mind-body dualism and modern science's teleological pursuit of anthropocentric progress. This distinction between mere cognition and deeply embodied knowledge becomes crucial for understanding how ceremonial practices constitute e/affective and enduring forms of environmental governance. "Such an ecosophy," Cajete writes, "would rebuild a unitary view of the cosmos in which everything is interdependent and moved by creative energy, one that views the Earth and the universe with reverence and explores our essential relationships and responsibilities therein."[16] Such an ecosophy, collaboratively enacted, challenges hierarchical systems of land management practices while (re)naturalizing place-based and cyclic relationships of reciprocal tending.

Aligned with this call for a synthesis of worldviews, Native Hawaiian scholar Manulani Aluli Meyer calls for "critical consciousness and respect for other ways of knowing," particularly for how "practices-in-place have developed a knowledge ethic shaped by the needs of place and people." She asserts "three ways in which to view and experience knowledge"—the objective/the subjective/and the quantum or spiritual—an "epistemologic trilogy" of mind/body/spirit that is not linear but "happening simultaneously and holographically." Recent findings in quantum physics, she argues, reveal a "(k)new understanding of the philosophy of knowledge inclusive of all three aspects of nature."[17] To know, then, might also convey an acceptance of meaning, beyond positivist facticity, that "builds on our innate sense of awe" and includes dimensions of ancestral memory, intuitive trust, and unbound love.[18] In *The Science of the Sacred*, Nicole Redvers (Dene) notes that "most of us are not given the opportunity to understand or develop the deep connectedness realized on a particle level."[19]

This willingness to decenter the human and to rediscover our place of belonging within a vast and vibrational web of ecological entanglements can be nourished by deep listening, including and especially listening to stories that expose the relatively recent and wholly artificial separation of humans and the environment. Nishnaabeg artist-activist Leanne Betasomasake Simpson notes that stories can propel us "to rebel against the permanence of settler colonial reality and to not just 'dream alternative realities' but to create them."[20] Stories, then, can "bring into being."[21] With attention to stories as alive and generative, existing and continuing beyond constructed constraints of time, much like ceremonies themselves, I turn to a literary work that weaves many stories—both in the oral tradition and with the characteristics of postmodern fiction—into a novel exemplifying this "(k)new understanding" of holistic interdependence: Leslie Marmon Silko's *Ceremony*.

Structured as a healing ceremony itself, the novel presents a sustained critique of settler-colonial land management—from eco-social assimilationist engineering to barbed-wire barbed-write claiming to uranium ore mining—while offering an alternative understanding of environmental stewardship figured through the protagonist, Tayo. While Silko confronts both the epistemic and material violence of colonialism and its "characteristics of dismemberment"—to borrow a powerful coinage from Thiong'o—she also underscores the power of ceremony to enable a re-membering, a reconnection of fragmented relationships across time and space in webs of multispecies kin.[22] As Tayo participates in ceremonial practices connecting dynamic and trans-historical worlds, Silko reveals how Indigenous "management"—what I will hereafter refer to as stewardship in Indigenous contexts—operates through deeply embodied relational knowledge rather than extractive control. Through patience and deep listening, stories and ceremonies become tools of renewal that counter the dismembering violence of settler colonialism while nourishing sensory attunement and a sense of belonging with more-than-human worlds. This expansive eco-ontological orientation reveals that land management/stewardship is always-already relational, a system of vibrations with endless reverberations. Whether or not these relations are visible to the human eye, they are real, and always aural—audible not only in what sounds but in what falls silent, in the generative tension between presence and absence.

Despite abundant scholarship on *Ceremony*, none offer sustained attention to Silko's treatment of aurality. The aural here includes both the ear and the aura, a "subtle emanation" that originates in the Greek word for "breath (of air)."[23] Breathing and winds, the dynamic and sentient breath of the earth, figure prominently in Silko's novel. I am interested in reading sonic encounters—including silence and listening—as not only barometers of health but also agents of epistemic and relational transition within interconnected ecological communities. *Ceremony* illustrates the potential of what Steven Feld calls *acoustemology*, a kind of sound-based knowing "grounded in the basic assumption that life is shared with others-in-relation" that invites us to "attend and attune."[24] This acoustic dimension of environmental stewardship nourishes an appreciation for how ceremonial practices constitute not merely cultural but also scientific renewal that challenges and resists colonial territorial control. Thus, sonic encounters are not only epistemic but also embodied, affective experiences that relate and restore.

Resonance, Return, and Re-Membering in *Ceremony*

First published in 1977, *Ceremony* was critically acclaimed and became "one of the most frequently taught contemporary novels in higher education."[25] The novel invites readers to enter a seemingly fragmented story-world that resists Euro-Western notions of linear time with ecologically cyclic and cosmological orientations. Silko, of mixed ancestry like her most heroic characters, dissolves racial and cultural boundaries. She incorporates the Laguna Pueblo five worlds' cosmology while weaving in elements from the Diné Red Antway ceremony and Hopi kivas and katsinas.[26] The novel immerses readers in these plural and permeable worlds before following Tayo's healing journey in the fifth, the current world.

The spiritual beings in *Ceremony* interact with and through humans, becoming, at times, indistinguishable from them. They "are not metaphors used to imply a 'holistic system' of ecological values," writes Louis Owens, but a real and simultaneous "part of the reality in which Tayo is subsumed."[27] They are both more-than-human and literal, already ontologically real for some; for others, they reactivate what Cajete calls "the metaphoric or nature mind," a source of knowing that "slowly recedes into the subconscious" as the rational mind develops.[28] This nature mind expands epistemological capacity beyond the merely cognitive and provides a pathway toward literal perceptions and relations to the world(s) as Indigenous epistemologies understand it/them. "Native community," writes Cajete," is about living a 'symbiotic' life in the context of a 'symbolic' culture, which includes the natural world as a vital participant and co-creator of community."[29]

Set largely in the drought-prone North American Southwest, the central plot involves Tayo, born of a Laguna mother and unknown (or undisclosed) man, from whom he attributes his blue eyes. He navigates compounded traumas, including his mother's abandonment, the boarding school system, and prisoner-of-war experiences—all manifestations of settler-colonial violence that alienated him from Laguna culture and the more-than-human world. As Paula Gunn Allen (Laguna Pueblo) notes, Tayo's illness "is a result of separation from the ancient unity of person with land, and his healing is a result of his recognition of this oneness."[30] The opening sequence situates readers not in place but *as* place becoming, beginning with Ts'its'tsi'nako or, Thought-Woman (also Spider Woman), crafting the story-world as she imagines it, thereby positioning readers as active participants who open to implication in this storying of the world. After this simultaneously unsettling and situating start, with ambiguous pronouns and unknown referents, Silko writes a single word—"sunrise"—providing an ecological point of collection, continuity, and return. Though initially silent and alone on the page, sunrise becomes a sacred, cyclic event that recurs throughout the novel; it also concludes the novel for a unified sense of continuity. Each return to sunrise, some through flashbacks that further disrupt a linear chronology, invites collective ecological connection and interdependence. Through songs, sacred silence, and ceremonial sounds, Tayo returns to health and "resonance"—what Cajete describes as "*relationship*, originally grounded on and inspired by people's feelings for the night sky and the cosmos." This resonance, a remembering that "everything has its source in the natural world," enacts a re-membering that nurtures connections to "the sources of one's life in all senses of the term."[31]

Tayo's plot begins with a vivid description of his feverish trauma after returning to the United States, having not only having survived World War II but also having witnessed the violent death of his cousin Rocky during the Bataan Death March in the Philippines, setting the visceral tone that Silko sustains throughout the novel. We are immediately submerged in unsettling dissonance. Tossing about in bed, Tayo hears the coiled springs squeaking as loud voices wrestle within him, from a man singing a familiar Spanish love song *Y Volverė* ("And I Will Return") to Japanese voices to familial Laguna voices, including an auditory fragment of his Uncle Josiah bringing him fever medicine as a child—then, back to Japanese soldiers shouting at him, then women, then a language unintelligible, perhaps his mother, until "all the voices were drowned in loud loud music from a big juke box."[32] While several note how Silko's description mimics the traumatized mind, it also illustrates what Chickasaw scholar Jodi Byrd terms "the cacophonies of colonialism," a coinage inspired by a Chickasaw/Choctaw term describing creative and destructive chaos at the edge of the Upper and Lower Worlds of creation. She defines it as "the cacophonous discourses which represent *and* suppress ongoing colonization."[33] The dissonance in this internal struggle includes echoes of hope and love, but they are fleeting and disjointed among the memories of fear, shame, and despair. While the immediacy of this scene reveals the sonic contours of Tayo's trauma, this recurring polyvocal experience signals an always-dynamic plurality of connections and continuities, both planetary and cosmological in scope.

Similar scenes depicting phonic cacophony in Tayo's consciousness punctuate the first portion of the novel, as do memories of "Indian School" and the shame it conveyed through its reduction of Indigenous Knowledge Systems to "old-time superstition."[34] These traumas recur in reactions to "sounds [that] took him back to the moment" and further sever his connection to the more-than-human world.[35] The sound of crushing seeds, for example, once a playful memory, "made him sick," taking him back to the crushing of Rocky's skull during the death march—the sound of wanton human-contrived destruction. Silko follows this dismembering trauma with a flashback to canyon springs during an earlier drought, "with the sound of water trickling" as Tayo's uncle Josiah reminds him of the ecological consequences of human alienation from homelands: "This is where we come from, see. This earth keeps us going. . . . [T]he wind and the dust, they are part of it, too, like the sun and the sky. . . . [D]roughts happen when people forget, when people misbehave."[36] Ecological sounds evoke memories, and Silko's positioning of these two sound experiences—one destructive and one instructive—invites consideration of our own ways of story-ing the present through sonic memories. Meyer writes, "Mind matters. It is the inside agency of body and the connecting essence of all that is material."[37] Despite the stark contrast in affect evoked by these memories, both reveal the relational impacts of listening, with the latter revealing the potential to hear with what Jo-Ann Archibald (Stó:lō) calls the ear "that is in our heart."[38]

Among all the sounds that flow through Tayo—mechanical, musical, natural, traumatic, healing—the voice emerges as a powerful aural force for relating to the animate world. Songs, for example, figure prominently as creative agents of "expression and perception" throughout the novel.[39] Simon Ortiz (Acoma Pueblo) notes that

> the substance of a song is emotional, but beyond that, spiritual, and it's real and you are present in, you are part of it. . . . A song is made substantial by its context—that is its reality, both that which is there and what is brought about by the song.[40]

When Rocky dies during the Bataan Death March, with the pounding sound of jungle rain surrounding them, Tayo "damned the rain until the words were a chant, and he sang it while he crawled upon the mud." During this traumatic dissociation, Tayo "could hear his own voice praying against the rain" which subsequently leads to years of drought on the Laguna Pueblo Reservation.[41] Immediately following the rain-cursing scene, Silko shares a Laguna story describing the scolding of Reed Woman, who was always bathing, by her sister, Corn Woman, who was always working. The ensuing drought in the simultaneous stories further entwines the plural worlds. The connection of ecological and human trauma across time and space provides a through line in the novel, merging interior and exterior landscapes while serving as a reminder that "matter is not separate from spirit."[42] Tayo's curse, a marker of personal loss, measurable in multiple registers, has multidimensional ramifications. He has forgotten that "a realm of unseen connection patterns exist and we are the causal linkages that alter its capacity."[43] Yet, Tayo is loving, a quality demonstrated by his sense of responsibility and compassionate treatment of both human and nonhuman relatives. His initially painful sensitivity signals his expansive and immeasurable capacity for both giving and receiving love.

Tayo's love, a biophilia weakened by traumas but returned through the healing of ceremony, can resist the evil forces, what Silko refers to as "witchery."[44] This witchery predicts that humans "will grow away from the earth/ then they grow away from the sun/then they grow away from the plants and animals."[45] Though tangled in violent traumas, his instinctual memory of this greater and inclusive connectedness remains. He remembers—a kind of knowing rooted in resonance—that elemental sounds like the wind provide knowledge for those who listen. In a flashback to an earlier drought, just before the war, Tayo goes back to that narrow canyon, listening to the water flowing into a pool, remembering how, during a drought, "the holy men listened to the winds at dawn" before telling people it "was time to dance for rain." While Silko does not reveal the ritual, she describes its curative effects as Tayo achieves a sense of interdependence, of resonance. In a powerful scene depicting his awareness of interconnected webs of relations, his memories empower and heal: "The things he did seemed right, as he imagined with his heart the rituals the cloud priests performed during a drought."[46] Here, he remembers the truth of Laguna stories that "human beings could understand what the animals said," and he "saw a world made of stories. . . . It was a world alive, always changing and moving."[47] This knowing, grounded in animate epistemology, challenges mechanistic views of human dominance and opens space for what Kimmerer calls "a grammar of animacy," a way of living in "a world with a democracy of species, not a tyranny of one."[48] Deep Listening nurtures resonance, a re-membering that restores relationships across the constructed divisions imposed by settler-colonial ideologies.

The gift of rain that follows precipitates Tayo's spiritual-erotic-ecological encounter with a "feminine life force of the universe" named Night Swan, a Mexican woman who "was like the rain and the wind; age had no relation to her."[49] These hyper-sonic scenes are vibrational in their dissolution, renewal, and union. Tayo could "hear his heart beating" as he drives to her home, with the prayed-for raindrops "splattering" and "rattling." Upon arrival, he hears *Y Volvere* on her victrola, and "the room pulsed with feeling."[50] Once "the music had stopped, and there was only the sound of the storm," the sonic agents enmesh bodies of all kinds:

> She whispered in Spanish. . . . [H]er sounds were gentle and the storm outside was loud. He could hear the rain rattling the roof and the sound of the old cottonwood tree straining in the wind. . . . She breathed harder and he breathed with the same rhythm. She moved under him, her rhythm merging into the sound of the wind shaking the rafters and the sound of the rain in the tree. And he was lost somewhere, deep beneath the surface of his own body and consciousness.[51]

The wind and rain, as agents of constant change and continuity, become sonic elements of a (k)new entanglement. "You do not have to understand what is happening," she tells him. "But remember this day. You will recognize it later."[52]

This deeply embodied spiritual-erotic-ecological encounter does indeed remain with Tayo, and it guides him through his search for healing after the war and during the drought he cursed into being. His experiences with a "medicine man" named Ku'oosh are only partly successful, as Ku'oosh represents the older, largely unchanged Laguna ways. It is the Diné medicine man, Betonie, connected to Tayo through Ku-oosh, who understands that "things which don't shift and change are dead things."[53] The mixed ancestry of Silko, Tayo, Night Swan, and Betonie underscores "the long history of incorporating into Laguna tradition what was brought in from other groups, so many stories emphasize continuing change."[54] Thus, Betonie understands "that it became necessary to create new ceremonies." As the ceremony begins, Betonie explains, "This has been going on for a long long time. . . . We all have been waiting for help a long time. But it never has been easy. The people must do it. You must do it." And so, with the ancestral knowledge of many voices guiding the creation of a new ceremony, we learn that Tayo's "cure would be found only in something great and inclusive of everything."[55] As Tayo seeks healing that exceeds the self, we learn that the stakes are cosmological in proportion, a fact illustrated by Betonie's tracing of constellations in the sand that later lead Tayo to recognize the patterns of belonging in the night skies.

The curative power of song and its creative agency come to the fore during the second night of ceremony, atop the Chuska Mountains. As Betonie sings the ritual into being, walking Tayo through bear footprints and sacred hoops, he sings of return: "I will bring you through my hoop. I will bring you home." Flowing from future into present tense, the singer merges with sacred elements, evoking the unification of human *as* nature in a recursive sequence of gathering, confluence, offering, and emergence: "I have the dew,/a sunray falls from me,/I was born from the mountain/I leave a path of wildflowers/A raindrop falls from me/I'm walking home/I'm walking back to belonging."[56] This ceremonial song illustrates how, as relational practices, "functional ontologies operate through Indigenous song" and actively reconstitute Indigenous ways of being and knowing through embodied participation.[57] Still, the power of "witchery"—a destructive force that both transcends and conscripts racial difference—is strong, and while Tayo's health improves, the slow work of deep, collective change remains.

But—he is on the path to healing, to re-membering his belonging among a multispecies commons, and he follows the stars to the home of T'seh, a mountain spirit whom Gunn Allen describes as the "creatrix of waters of love that flow from a woman and bless the earth and the beloved with healing."[58] He hears her voice before he sees her, and a blowing wind makes speech difficult until they go inside, where the flames of the fireplace "snap the kindling and suck at the wind that came down the chimney." Surrounded by these elemental agents of creation and change, their spiritual-erotic-ecological encounter reveals Tayo's return to belonging and restored trust in himself, his capacity to love. "Afraid of being lost," he imagines his body into a landscape, "repeat[ing] trail markers to himself. . . . [T]his is my voice calling out to her."[59] This encounter reveals repair in terms both corporeal and more than-human; intimacy enacts a re-membering, a returning of awareness, kinships, and obligations "into the fleshly present."[60]

As he re-members, Tayo recognizes and recovers the voice he had lost in the jungle rain, and as "dawn woke the instinct in the dim memory of the blood," he sings a song to the sunrise the next morning.[61] This sacred encounter illustrates the dual meanings of "becoming sound" that Dustin Tahmekerahe (Comanche) describes as the "interdependent relations between emitting sound as the formations of sonic vibrations in the air and becoming sound as a method toward restoring good health through cultural ways of listening and healing."[62] Silko infuses Tayo's song with ancestral-instinctual resonance:

> He repeated the words as he remembered them, not sure if they were the right ones, but feeling they were right, feeling the instant of the dawn was an event which in a single moment gathered all things together—the last stars, the mountaintops, the clouds, and the winds celebrated this coming. The power of each day spilled over the hills in great silence.

Karen Barad's insights on memory as "not merely a subjective capacity of the human mind" but actually "written into the worlding of the world" suggest the planetary scope of this re-membering.[63] The silence, too, is a call to slow down, to listen, to unite with and belong to/with place beyond the limits of language. In this scene of expansion and becoming, we also sense alignment with Cajete's "modern ecosophy," wherein Tayo recognizes "a unitary view of the cosmos in which everything is interdependent and moved by creative energy. [he] views the Earth and the universe with reverence." As he re-members within "the implicate order" of a shared ecological and cosmological commons, Tayo still needs to "explore [his] essential relationships and responsibilities therein."[64]

The ceremony continues as Tayo leaves T'seh to search for Josiah's Mexican spotted cattle, hardy "descendants of generations of desert cattle, born in dry sand and scrubby mesquite." When he locates them through the barbed wire of a sprawling ranch, the journey proves perilous: Tayo is spotted by ranchers and falls unconscious when his mare stumbles, leaving him wounded and captive as a snowstorm approaches. Though his pain leads him to consider "sinking into the elemental arms of mountain silence," he endures—and the more-than-human world aids him: both a mountain lion that lures the ranchers away and a snowstorm that protects him. After this reminder of resonance and resilience, Tayo smiles, finding piñons to eat, and "crack[s] the shells one by one." He encounters a hunter "singing a chant of recognized phrases." Yet, the song "wasn't a Laguna song; it sounded like a Jemez song or maybe one from Zuni."[65] When Tayo and the hunter return, to find the spotted cattle already safely corralled with T'seh, she explains that they are "just like runoff after a rainstorm," so you build a trap that follows "the continuation of the arroyo."[66] In this stage of ceremonial healing, Tayo increasingly mimics his nonhuman relatives, "imitating the gentleness of the bees" as he uses a blue feather to spread pollen to "flowers that had no bees." In these acts of attunement and reciprocal care, Tayo practices the ecorelational literacy that T'seh embodies.

T'seh also represents the potential in rematriation—"a call to reestablish Indigenous landscapes, bring back Indigenous traditional ecological knowledges and to give stolen power back to the feminine."[67] While spending a dream-like and intimate summer with her as they tend the spotted cattle, Tayo absorbs not only her plant and seed knowledge but also her way of being in relation with the land—a practice of loving attentiveness that weaves together observation, listening, patience, and reciprocity. When she leaves him, instructing him to "remember everything," Tayo appears healed; when he sees the sunrise, "it was all so beautiful, everything, from all directions, evenly, perfectly, balancing day with night, summer months with winter. . . . The strength came from here, from this feeling. It had always been there."[68]

However, the ceremony remains incomplete until he finds himself in an abandoned uranium mine near the Trinity Site, surrounded by the ecological wounds of mismanagement and extractive machinations of modern warfare. There he recognizes the unity of patterns that dissolves false divisions: "he recognized why the Japanese voices had merged with Laguna voices, with Josiah's voice and Rocky's voice," all cultures were converging in the middle of witchery's final ceremonial sand painting. They were all "united by the fate the destroyers planned for all of them, for all living things." The stone "streaked with powdery yellow uranium, bright and alive as pollen" reveals the perverse mimicry of destruction: "they had taken these beautiful rocks

69 Silko, *Ceremony*, 246, emphasis added.
70 Silko, *Ceremony*, 255.
71 Wilson and Hughes, "Why Research is Reconciliation," 11.
72 Shiva, *Reclaiming the Commons*, 249.
73 Robinson, *Hungry Listening*, 15.
74 Redvers, et al., "The Determinants of Planetary Health," 157.
75 Bioneers, "Landback: Restoring People, Place and Purpose."
76 Cook, "Amnesty International Calls Greenpeace Loss a 'Chilling Verdict'," 1.
77 Wildcat, *Red Alert! Saving the Planet with Indigenous Knowledge*, 127.

Bibliography

Allen, Paula Gunn. "The Psychological Landscape of *Ceremony*." *American Indian Quarterly* 5, no. 1 (1979): 7–12.

Allen, Paula Gunn. *The Sacred Hoop: Recovering the Feminine in American Indian Traditions*. Beacon Press, 1992.

Barad, Karen. *Meeting the Universe Halfway: Quantum Physics and the Entanglement of Matter and Meaning*. Duke University Press, 2007.

Barad, Karen. "Troubling Time/s and Ecologies of Nothingness: Re-Turning, Re-Membering, and Facing the Incalculable." In *Eco-Deconstruction: Derrida and Environmental Philosophy*, edited by Phillipe Lunes and David Wood, 206–248. Fordham University Press, 2018.

Bell, Robert. "Circular Design in *Ceremony*." In *Ceremony: A Casebook*, edited by Allen Chavkin, 23–39. Oxford University Press, 2002.

Bioneers. "Landback: Restoring People, Place and Purpose." July 2023. https://bioneers.org/landback-restoring-people-place-and-purpose-zmbz2307/.

Byrd, Jodi A. *The Transit of Empire: Indigenous Critiques of Colonialism*. University of Minnesota Press, 2011.

Cajete, Gregory. *Native Science: Natural Laws of Interdependence*. Clear Light Publishers, 2000.

Chavkin, Allen. "Introduction." In *Leslie Marmon Silko's Ceremony: A Casebook*. Oxford University Press, 2002.

Colorado River Compact. Santa Fe, New Mexico: U.S. Bureau of Reclamation, November 24, 1922. Accessed February 2, 2025. www.usbr.gov/lc/region.

Cook, Marnie. "Amnesty International Calls Greenpeace Loss a 'Chilling Verdict'." *Native Sun News Today*, March 26–April 1, 2025.

Environmental Protection Agency. "Abandoned Mines Cleanup." *Environmental Protection Agency*, 2025. https://www.epa.gov/navajo-nation-uranium-cleanup/aum-cleanup.

Evans, Simon. "Analysis: Which Countries Are Historically Responsible for Climate Change." *Carbon Brief*, 2021.

Feld, Steven. "Acoustemology." In *Keywords in Sound*, edited by David Nowak and Matt Sakakeeny. Duke University Press, 2015.

Haraway, Donna. *Staying With the Trouble: Making Kin in the Chthulucene*. Duke University Press, 2016.

Howard, Laniikque. "Addressing Water and Wastewater Challenges in Tribal Nations." *Administration for Children and Families*, August 25, 2022.

Hu, Sheila. "The Dakota Access Pipeline: What You Need to Know." *National Resources Defense Council*, June 12, 2024.

Intergovernmental Panel on Climate Change (IPCC). *Climate Change 2023: Synthesis Report: Summary for Policymakers*, edited by Core Writing Team and led by Hoesung Lee. Geneva: IPCC, 2023. https://www.ipcc.ch/report/ar6/syr/

Justice, Daniel Heath. *Why Indigenous Literatures Matter*. Wilfrid Laurier University Press, 2018.

Kimmerer, Robin Wall. *Braiding Sweetgrass: Indigenous Wisdom, Scientific Knowledge, and the Teachings of Plants*. Milkweed Editions, 2013.
Kimmerer, Robin Wall. "On Choosing to Belong to a Place." *Common Good Collective*. https://commongood.cc/reader/on-choosing-to-belong-to-a-place/.
Meyer, Manulani Aluli. "Holographic Epistemology: Native Common Sense." *China Media Research* 9, no. 2 (2013): 94–101.
"Native American Water Rights Settlement Project." *UNM Digital Repository*. University of New Mexico. Accessed October 28, 2025. digitalrepository.unm.edu/nawrs/.
Nobiss, Sikowis. "Why 'ReMatriate' Is a More Inclusive Term for Returning Land to Indigenous Peoples." *Reckon*, October 3, 2023.
Oliveros, Pauline. "Quantum Listening: From Practice to Theory (To Practice Practice)." *Sound Art Archive at the Center for Deep Listening*, December 1999. https://www.deeplistening.rpi.edu/deep-listening/.
Ortiz, Simon. *Song, Poetry and Language—Expression and Perception*. Tsaile: Navajo Community College Press, 1977.
Owens, Louis. "Leslie Silko's Webs of Identity." In *Leslie Marmon Silko's Ceremony: A Casebook*, 91–116. Oxford University Press, 2002.
Oxford University Press. "Aura, n." *OED Online*. Accessed July 1, 2025. https://www.oed.com/dictionary/aura_n.
Redvers, Nicole. *The Science of the Sacred*. North Atlantic Books, 2019.
Redvers, Nicole, et al. "The Determinants of Planetary Health: An Indigenous Consensus Perspective." *The Lancet Planetary Health* 6, no. 2 (2022): 156–163.
Robinson, Dylan. *Hungry Listening: Resonant Theory for Indigenous Sound Studies*. University of Minnesota Press, 2020.
Ruoff, Lavonne A. "Ritual and Renewal: Keres Traditions in the Short Fiction of Leslie Silko." *Multi-Ethnic Literatures of the United States* 5, no. 4 (Winter 1978): 2–17.
Shiva, Vandana. *Reclaiming the Commons: Biodiversity, Indigenous Knowledge, and the Rights of Mother Earth*. Gabriola Island: New Society Publishers, 2020.
Silko, Leslie Marmon. *Ceremony*. Viking Press, 1977.
Simpson, Leanne Betasamosake. "Land as Pedagogy: Nishnaabeg Intelligence and Rebellious Transformation." *Decolonization: Indigeneity, Education, and Society* 3, no. 3 (2014): 1–25.
Tahmahkera, Dustin. "Becoming Sound: Indigenous Soundscapes and Decolonial Sound Studies." *Sounding Out!* (blog), October 9, 2017.
Thiong'o, Ngũgĩ wa. *Something Torn and New: An African Renaissance*. Perseus Books Group, 2009.
Webb, Jeremy. "Historic Tribal Water Rights Settlement—Poised to Reshape Arizona's Future." *Economic Development Journal of Mohave Country*, April 2025.
Wildcat, Daniel R. *Red Alert! Saving the Planet with Indigenous Knowledge*. Fulcrum, 2009.
Wilson, Shawn. *Research is Ceremony: Indigenous Research Methods*. Fernwood Publishing, 2008.
Wilson, Shawn, and Margaret Hughes. "Why Research is Reconciliation." In *Research and Reconciliation: Unsettling Ways of Knowing Through Indigenous Relationships*, 5–19. Canadian Scholars, 2019.
Winters v. United States. 207 U.S. 564 (1908).

Part 4

Resourcing Love to Actualize Otherwise Worlds

17 Grassing Gettysburg

Management, Memory, and Meaning

Michael R. Barnes and Amy T. Hamilton

Grass

Pile the bodies high at Austerlitz and Waterloo.
Shovel them under and let me work—
I am the grass; I cover all.
And pile them high at Gettysburg
And pile them high at Ypres and Verdun.
Shovel them under and let me work.
Two years, ten years, and passengers ask the conductor:
What place is this?
Where are we now?
I am the grass.
Let me work.

—Carl Sandburg

On a sunny Saturday in August 2024, we arrived at Gettysburg National Military Park (GNMP) in Pennsylvania. There we joined long queues of visitors in tour buses, cars, SUVs, trucks, motorized three-wheelers, and Segways. Routing the official "Auto Tour" app through our rental car's speakers, we drove from site to site, following the history of the battle and imagining thousands of troops camping, marching, and fighting on the now peaceful landscape. Ensconced in the comfortable air-conditioned quiet of the car, we listened to the tour guide's version of the battle, craning our necks to find the sightlines he described and attempting to replace the tourists, monuments, and wayside parking lots with the sights and sounds of battle. Pulling over intermittently, we joined other visitors in their American flag T-shirts and reproduction Union kepis. We tried to count canons (and quickly gave up; there are hundreds) and commented on the sheer number of monuments dotting the landscape (more than 1,300) (see Figure 17.1).

After our three-hour tour, we stopped by Soldiers' Cemetery to visit the site of Lincoln's "Gettysburg Address" and the semicircle of grave markers, most of them unnamed, which indicate the final resting place of over 3,500 of

DOI: 10.4324/9781003581185-22

Figure 17.1 A cannon, monument, and brome grasses: a scene replicated hundreds of times across the site.

Source: Photograph by Amy T. Hamilton

the Union soldiers killed between July 1 and July 3, 1863. Weaving between gravestones and tour groups, we considered the placement of the cemetery and searched for the stones honoring the soldiers from our states (Minnesota and Michigan) who rest there. Finally, we stopped by the visitor center and gift shop, perusing the displays of photographs and the scores of books, souvenir shot glasses, and personalized keychains for sale.

The Park offers visitors an almost overwhelming glut of information: the tour is an exhaustive recounting of each hour of those three days of fighting, as well as the days leading up to them as troops amassed in the area. The tour aims to maximize and standardize the experience of the site in a similar manner as described by Eytan Pol who writes about the impacts of hiking apps on wilderness experiences in this collection. The monuments note the participation of each division from both the Army of the Potomac and the Army of Northern Virginia, each state represented, each important general, and each decisive moment in the battle. While movement through the park is highly structured with tours, trails, roads, markers, monuments, and signage, the number of visitors and their individual movements resist such containment and, perhaps, are closer to the history of orchestrated chaos that the site has seen: cars move like cavalry charging through the winding roads of the site, the rumble of buses recalls artillery heard from far and wide, and tour groups march like soldiers. But within the commotion and info-whelm, there are notable silences: shockingly little about the context of the war (slavery), even less about Indigenous tenure on the land (historical or current), and next to nothing about the land itself (a notable silence from the NPS). As a result of these silences, the battlefield, a clearly *placed* site with a history that is wholly dependent on US colonialism, is oddly dematerialized and decontextualized. The NPS's internal documents, however, reveal an intense focus on land and vegetation, which is not reflected in the Park's public-facing publications, tours, or signage. Land management is of paramount importance to the preservation and continued care and development of the Park, but visitors have very few opportunities to engage with that narrative via park interpretation.

From its inception, NPS management of Gettysburg has been deeply linked to the creation and maintenance of particular ideas about memory, patriotism, and history. And, crucially, one of the most important areas of concern is the management and curation of grass on both the battlefield and the cemetery. As the official park website states:

> Vegetation management is an important responsibility of the park's natural resource staff. The primary goals of Natural Resource Planning at Gettysburg . . . are to (1) restore and perpetuate the battlefield as it appeared at the time of the Battle of Gettysburg in July 1863 and to (2) preserve resident fauna and flora that are compatible with the goal of historic accuracy. With these goals, park personnel conduct floral inventories, monitor seedling recruitment, and map vegetative cover types. Vegetation management is also a critical part of the park's landscape rehabilitation plan.[1]

Further, park staff actively "work to combat several invasive plant species" through a combination of "chemical methods, mechanical methods, hand pulling and sprays."[2] The function of this management is to create the appearance of a pristine expanse—significantly aiming for a recreation of the landscape as it was in 1863—a place where visitors can imagine the drama of the battlefield and the reconciliation of the nation. The extensive and intensive management of the park demonstrates that grass management activities are—or are intended to be—hidden to visitors by the grasslands themselves.

In *Memorials Matter: Emotion, Environment, and Public Memory at American Historical Sites*, Jennifer Ladino explains that the NPS has a "huge responsibility for narrating the intense history of the U.S., and so, for managing the relationship between public memory and national identity."[3] As visitors, she continues, "our bodies are carefully managed along with natural resources NPS sites celebrate and enclose."[4] At Gettysburg, our bodies are directed to move along the road and stop at particular lookouts in an attempt to align our embodied experience with the material landscape and historical events. We, like the managed grass, are uneasily placed so that our experience is fractured in space and time; we look out over a contemporary landscape that is manipulated to look like a historical landscape as we listen to a recording from the Gettysburg website that asks us to imagine events from 160 years ago.

The complexity of a site like Gettysburg calls urgently for interdisciplinary analysis. Just as the landscape and our responses to it resist enclosure and control by NPS management, so too do they exceed the limits of individual disciplines. Indeed, Jonathan Cane's contention that "landscapes should be considered 'verbs', unfinished processes, inconclusive attempts at fixing a permanent vision of nature, rather than as 'nouns', which can be surveyed, owned or possessed"[5] resonates with the need for interdisciplinary approaches that can consider matter and meaning from a variety of perspectives, as we do in this chapter.

Citing Bruno Latour, Stacy Alaimo in *Bodily Natures: Science, Environment, and the Material Self* argues that the pressing climate questions confronting us require a reknitting of "scientific, sociological, and textual knowledge practices."[6] Similarly, in *Slow Violence and the Environmentalism of the Poor*, Rob Nixon offers the concept of the "ecotone" where "adjacent communities of vegetation . . . meet and interact" as a way to consider the importance of different disciplines collaborating to create potentially "dynamic combinations."[7] However, as Alaimo notes, such interdisciplinarity "is surely no small feat, given that scholars are trained, for the most part, to engage in only one of these three modes of investigation."[8] Academic training touches on all parts of the scholarly process, from framing research questions, to researching, to writing and understanding findings. Scholars committed to interdisciplinary work face a difficult challenge of articulating the internalized and therefore unconsciously routinized practices of their own discipline while simultaneously learning the unfamiliar and sometimes opposing practices of another. These challenges can lead even those who truly believe in the potential of interdisciplinary work to shy away. And yet talking across and through disciplinary differences is vital if we are going to responsibly understand and address climate crises.

The following chapter responds to Alaimo, LaTour, and Nixon's call for interdisciplinary collaboration. As a study co-written by a literary scholar with expertise in ecocriticism and a social scientist who specializes in turfgrass lawns, this chapter proposes an interdisciplinary approach that synthesizes diverse disciplinary methodologies. Reading both management policies and land as polyvocal texts, we interrogate how the selection and cultivation of grasses in two distinct landscapes at Gettysburg (battlefield and cemetery) participate in the symbolic and material creation of meaning, memory, and identity, and how the grass itself resists containment and management, pointing to other meanings and other histories. Grass—that ubiquitous plant that somehow manages to encourage extreme scrutiny (e.g., the American lawn) and benign disregard (e.g., grass-covered fields)—reveals how land management functions at levels of intense detail, down to blades of grass, while also focusing attention on larger impacts (e.g., biodiversity, greenhouse gas emissions). Our interdisciplinary approach offers enriched avenues for framing the possibilities of management and reimagining grass as not an unyielding monoculture but rather as a tapestry that is symbolically and materially composed of many shades of green.

Grassing the Past: History of Land Management at Gettysburg

The federal history of land management at Gettysburg began only months after the battle itself with the dedication of the national cemetery. President Abraham Lincoln traveled to the site in November of 1863, while the Civil War still raged. There he gave one of the defining speeches of his presidency, indeed of *any* US presidency, known familiarly as "The Gettysburg Address." The speech marked the dedication of the Gettysburg National Cemetery, the beginning national management of the site before the end of the war. In the years since, the management of both cemetery and battle site has gone through many phases and changes, permutations that reflect shifting national perspectives on memory, land, preservation, and identity.

The first steps in creating the Gettysburg National Cemetery were taken in response to a clear need that followed in the wake of the catastrophic battle. The clash at Gettysburg lasted only for three days, yet the devastation was crushing: approximately 51,000 casualties, "including 10,000 killed, over 30,000 wounded, and nearly 11,000 missing or prisoners of war." Seven thousand of those fallen soldiers "lay haphazardly buried in shallow graves on the fields where they fell."[9] These shallow graves presented serious aesthetic and health-related concerns for the citizens of Gettysburg who quickly moved to establish a cemetery for the soldiers adjacent to the existing Evergreen Cemetery. In the midst of the ongoing war, white Union soldiers were moved to this new site on Cemetery Hill, while the Confederate soldiers remained where they fell until they were removed to southern cemeteries in 1870.[10] Significantly, no Black soldiers officially fought at Gettysburg, and the cemetery remained segregated until 1885. Today, only two Black veterans of the Civil War are known to be buried there.[11]

Mere weeks after the battle, a movement was underway to preserve the battlefield itself. Local residents quickly started to work toward commemorating the site, and within a year, the Pennsylvania legislature created the Gettysburg Battlefield Memorial Association, "the nation's earliest attempt to preserve Civil War battlefields."[12] Indeed, as early as July 24, 1863, "the *New York Herald* predicted that Gettysburg 'will, in the future, be one of the nation's altars.'"[13]

In the 160 years that have followed these first acts of commemoration, the management of both cemetery and battlefield has changed and developed as national ideas about patriotism and the significance of the Civil War have emerged. Historian Jim Weeks avers that these changes reveal "the apparent paradox between Gettysburg as both a site of commemoration and the object of commerce, and the evolution of both memory and tourism."[14] *What*

Gettysburg means is intimately bound up in *how* that meaning is created, communicated, and commodified.

Unsurprisingly, the meaning of the conjoined sites of the cemetery and battlefield diverged almost immediately following their establishment. Weeks quotes William Saunders, the landscape architect who designed the cemetery, as hoping that the site would "realize a pleasing landscape and pleasure ground effect," ideal for contemplative strolls. Further, Saunders also designed the "small, uniform gravestones" in an effort to ensure that visitors would be able to focus on the awe-inspiring views from Cemetery Hill.[15] In other words, the early focus at the cemetery, which has largely been maintained, was on creating a space that would evoke peace and contemplation, a site for considering the sacred sacrifice of the Union soldiers. In contrast, the battlefield has alternately figured as a site denoting patriotic pride, adventure, and valor. As historian Thomas A. Desjardins explains, "American mythology has established Gettysburg as the greatest, biggest, most important, most heroic, most savage, bloodiest battle the nation ever fought. . . . It is our Waterloo, our Stalingrad, and our defining moment."[16]

In order to realize that "defining moment," early management efforts at Gettysburg focused on rehabilitating the torn ground that was littered with munitions and bodies. In 1865, when John Townsend Trowbridge toured Gettysburg, a local man told him, "The ground was all covered with dead horses, and broken wagons, and pieces of shells, and battered muskets, and everything of the kind, not to mention the heaps of dead."[17] While the bodies had been buried by the time of Trowbridge's visit, the battlefield itself was still littered with

> soldiers' devices, hastily adopted for defense. A rude embankment of stakes and logs and stones, covered with earth . . .[and] breastworks, as it were, consisting of rocks heaped up by the trunk of a tree or beside a larger rock, or across a cleft in the rocks, where some sharpshooter stood and exercised his skill.[18]

Yet, even in these early descriptions, Trowbridge struggled to reconcile stark reminders of wartime violence with a desire for serenity. He writes of a copse of trees as

> very beautiful . . . with their great, silent trunks, all so friendly, their clear vistas and sun-spotted spaces. Beneath reposed huge, sleepy ledges and boulders, their broad backs covered with lichens and old moss. A more fitting spot for a picnic, one would say, than for a battle.[19]

Trowbridge's impulse to aestheticize the landscape and smooth over horror with rest and recreation is complicated by his own language as he imagines the boulders as bodies, eerily reminiscent of the soldiers who were killed.

Further confronting his move to peace and beauty are the trees themselves, which, as he notices, bear,

> more astonishing evidences of fierce fighting than anywhere else about Gettysburg. The trees in certain localities are all scarred, disfigured, and literally dying or dead from their wounds. The marks of balls in some of the trunks are countless. Here are limbs, and yonder are whole tree-tops, cut off by shells. Many of these trees have been hacked for lead, and chips containing bullets have been carried away for relics.[20]

The tension underlying Trowbridge's descriptions is one that continues to animate Gettysburg today, where a drive to create a space for tranquil contemplation is challenged by an intense interest in the violence and horror the site has witnessed.[21]

Since its founding in 1864 by the citizen-led Gettysburg Battlefield Memorial Association, the management of the Gettysburg battlefield site has transferred hands only twice.[22] After 31 years of civilian oversight of the battlefield and cemetery, the management of the site was transferred to the federal government, specifically the War Department, and was officially designated as a national military park.[23] The Gettysburg site was only the fourth national military park established in the country. In 1933, oversight was passed to the NPS, and until the 1990s, the site has been managed by a preservationist ethos.[24] The NPS's stated goal for the land it oversees is to "preserv[e] unimpaired the natural and cultural resources and values of the National Park System for the enjoyment, education, and inspiration of this and future generations."[25]

J. Christian Spielvogel argues that this preservationist perspective glosses over deep conflicts in the site's history in favor of a version of history that prioritizes "regional harmony and reconciliation between whites."[26] By the late 1990s, the NPS had realigned its focus in order to center slavery as the principal cause of the Civil War, a focus that required places like Gettysburg to shift away from the minutiae of battle strategy to considering the larger meaning of the site. Indeed, a sign that greets visitors to Gettysburg reads,

> The Civil War was fought over three issues—survival of the Union, the fate of slavery and . . . what it means to be an American. The war resolved the first two issues. The nation struggles with the third to this day.[27]

More generally, across its many sites, the NPS has committed itself to the truth that "national parks are native lands" and "tribal priorities, values, and stories" should be incorporated into park management "at every level."[28] Yet, even with this shifted NPS focus and with Gettysburg's particular history, slavery and Indigenous perspectives remain largely offstage in park interpretation and resource management.

In the midst of these silences and erasures lies the land itself. Gettysburg is a particularly complex site to manage, given the intertwined nature of natural and built landscape elements across both contemporary and historical contexts. For example, the dueling mandates of providing access to park visitors through multiple forms of recreational activities on the one hand while minimizing disruption and impacts on sensitive habitats and restoration sites within the park on the other illustrate this complexity. Many such human-nature conflicts exist in the context of the site due to the size and diversity of the park's natural and cultural resources.

Grassing the Present: Current Conditions

Current management trends at GNMP strive to attain a sense of naturalness that belies intensive intervention. As Speilvogel notes, NPS management and interpretation are "informed by influential commemorative traditions of the past that reflect deeply embedded ideologies about race, gender, nationalism, and war." And while "their influence is profound, these commemorative traditions do not assert themselves forcefully on the landscape; they remain largely quiet and undisturbed, a seemingly natural part of the landscape preserved to its wartime appearance."[29] The grass that grows on these sites is a significant contributor to this goal, providing a verdant backdrop that passively conveys peace and ease by tapping into visitors' familiarity and comfort with meadows and fields.

Spielvogel maintains,

> The physical landscape, more so than other forms of visual interpretation (with the exception of photography, perhaps), appears to reflect a past unmediated by the subjective forces of history and human agency. Yet battlefields are designated, after all, as battlefield parks, and the beautifully maintained battlefield grounds, while undoubtedly a source of the Park Service's popularity, do contribute to its overall depiction of combat.[30]

The mediation of sites like Gettysburg is a meticulous and carefully constructed process whereby each sloping hill, each framed vista, and each contemplative retreat is composed to encourage particular emotions and reflections.

The site contains an abundance of biodiversity across multiple ecosystem types with over 500 different plant species alone, the vast majority of which are native to the area.[31] While the park contains a variety of ecosystems, one of the most important are the grasslands covering over 3,000 acres within the park.[32] Grasses at Gettysburg are a mix of introduced and native species along with a blend of cool-season and warm-season species. Common species on the site include switchgrass, timothy, orchard grass, brome grasses, fescues, and bluegrasses, among others.[33] In addition to grassland areas, the park and cemetery are home to managed turfgrass areas. Turfgrasses are a specific subset of grass species which provide a consistent ground cover and can tolerate mowing and foot traffic.[34] Common cool-season turfgrass species found across both the park and cemetery include Kentucky bluegrass, perennial ryegrass, and fine leaf fescues. While cool-season grasses dominate, the site lies on the northern edge of the transition zone, an area in which both cool-season and warm-season grasses can persist, and as such warm-season grasses such as zoysiagrass can also be found. Warm-season grasses are likely to grow in abundance via intentional management and naturally as the transition zone moves north due to climate change.

The combination of vegetation selections and management techniques aligns with visitor preferences for more "natural" landscapes and acceptance of on-site invasive species control.[35] The function of this management is to create the appearance of an untouched field, a beautiful but neutral backdrop for historical imaginings. In this way, both management and vegetation are supposed to go unnoticed. In a previous chapter in this collection, Dominique Aurilla Vargas, writing about the US-Mexico borderlands, argues, "The infrastructures that lie beneath everyday human interactions are innumerable; however, they are—by design—systems with which humans have few intentional, conscious everyday relationships." However, these infrastructures cannot disappear completely, and while the intention at Gettysburg may be to present a pristine landscape, the site more so is an extreme version of an informal garden. As Twedt et al. explains, informal gardens are "hardly distinguishable from natural ecosystems and show little apparent sign of human intervention although they are carefully wrought works of art."[36]

The tension between the seemingly untouched landscape and the intensive work required to maintain it regularly pushes past boundaries into public perception; the park today is awash in signs of human intervention. On the subtle end, tire marks left in the vegetation that go off trail and into the more "natural" areas allude to some type of management taking place just outside the purview of visitors. More obvious are marked trails and roads leading to maintenance buildings only accessible to park staff, away from visitors' gaze as they move from stop to stop. The most overt display of active management are several signs placed at a popular stop displaying the restoration efforts undertaken by park staff. Signs show a "before restoration" image directly emphasizing the human-led intervention to "re-naturalize" that area of the park (see Figure 17.2). It seems, then, that the goal of hiding the management in favor of the creation of a "natural" landscape is not presently being achieved, but perhaps as Carl Sandburg suggests in the poem that begins our chapter, grass at Gettysburg "covers all" and will shift memory and meaning over time as the grass continues to grow. And, perhaps tellingly, during various encounters with park staff both at the visitors' center and out in the park, few articulated knowing about or were able to discuss vegetation management and instead pointed us back to NPS websites.

Figure 17.2 Signage and landscape at Little Round Top with goldenrod, vetch, and brome grass, reflecting a before and after restoration view.

Source: Photograph by Michael R. Barnes

Meanwhile, the meticulously controlled turfgrass at the cemetery is meant to be less suggestive of naturalness than of order and honor. The policies that determine the management of the cemetery lawn are detailed and specific, covering mowing and weed control among other activities. Mowing regimes are highly structured with policy dictating specific heights of cut, alternate directions of mowing, and clippings management. Even more specific rules exist related to how to cut the grass around headstones and monuments, with directions to keep mowing equipment "6 to 12 inches from upright headstones" and specifying the use of an anti-scalping device to minimize damage to the grass and gravestones.[37] Such intensive mowing procedures can have significant consequences for the sustainability of grassy areas with mowing often being the most energy-intensive form of maintenance.[38]

Despite these precise and detailed requirements being laid out line by line in the guidelines, the mowing practice in reality diverges significantly. The grass at the site was unsystematically cut at multiple lengths, varying from 1 to 5" in length, with the longer lengths existing around gravestones and tight clusters dotted around (see Figure 17.3). The landscape report also points out the significance of the appearance of the grass: "A well-kept sward, graceful shade trees, pretty shrubs and evergreens are of greater importance and contribute more to the beauty of the cemetery than a promiscuous distribution of flower beds."[39] The emphasis here is on orderliness and harmony, contrasting the stateliness of grass and evergreens with the riotous fertility of flowers. Here, again, however, the reality of the site stands in contrast with the policy as written. The site during our visit was quite disorderly, with not only various lengths of grass but also turf in various states of heat stress and dormancy, tilting the color palette from the expected green more toward yellow and brown.

Another way in which the site attempts to keep a neat and orderly appearance is illustrated through how the cemetery defines and manages weeds. A weed is defined in the official guidelines as "a plant out of place," growing where it's not wanted.[40] This definition underscores the arbitrariness of deciding which plants remain and which are rooted out. Cemetery policies establish that grass areas should be "generally weed free," whereby weeds can be present but cannot distract from the "visual attractiveness of the landscape."[41] Management of nuisance weeds relies on chemical controls through the use of herbicides, with specific chemical formulations and application methods specified. Herbicide use can have significant environmental

Figure 17.3 Various heights of Kentucky bluegrass and perennial ryegrasses protrude about the broader canopy and over the gravestones across the cemetery landscape.

Source: Photograph by Amy T. Hamilton

consequences related to impacting biodiversity, especially among pollinator species.[42] Again, the site as a living place did not meet this "generally weed free" mandate as directed by policy. In contrast, the weeds on the site were, at times, more prominent than the grass (see Figure 17.4). Clover, crabgrass, and dandelion (among others) were plentiful, creating their own "green" cover where the grass provided none. Taken together, rather than an attempt to recreate a particular kind of wildness as with the battlefield, the practices of grass maintenance in the cemetery attempt to reflect a patriotism that is associated with military control and precision; however, such aims aren't met in reality.

Attempts to carefully monitor and control the landscape continue to dominate NPS documents even as the grass itself resists such direction. The management of grasses at national heritage sites such as Gettysburg is meant to create a sense of bucolic natural freshness, but obscures the deliberate creation of particular meanings and affective memory experiences that are intimately bound up with national identity, the legacy of slavery and emancipation, and the relationship between past and present. Spielvogel explains the pastoral as a "middle landscape" that favors a rural rather than a wild or manicured

Figure 17.4 A mélange of weeds (e.g., clover, crabgrass, and dandelion) overtakes drought-stressed grasses (Kentucky bluegrass, perennial ryegrass) at the cemetery, creating a mosaic of colors and textures disobeying the strict regulations of precisely managed turfgrass.

Source: Photograph by Michael R. Barnes

idea."[43] Significantly, this aesthetic is linked to the imagined pre-war idyllic South "as a land inhabited by faithful slaves and free of the dehumanizing forces of modernization."[44] In her chapter in this volume, Mika Kennedy cites Thy Phu's observation that "Cultivation shares roots with notions of culture; agricultural cultivation becomes a means of cultivating citizenship." While "cultivation" has a very specific meaning in the field of agricultural science, tying it here to the broader historical and cultural contexts of Gettysburg underscores the complexity of NPS management decisions. And all of these considerations are further complicated by the United States' ecological imagination that positions grasslands, like those recreated at Gettysburg, as a symbol of freedom, progress, and American exceptionalism.

(Re)membering Grass at Gettysburg

Status quo narratives, such as those enacted and reinforced at GNMP, resist alternative imaginaries of grass as a more diverse and sustainable form of nature. They are examples of "landscape freezing" or attempting to arrest a landscape at one historical moment. Landscape freezing stands in contrast to ecological preservation which focuses on the reinstatement of a healthy ecological system which is inevitably expected to change.[45] Jennifer Murray argues,

> Opponents to "landscape freezing" maintain that it leads to "purposeful amnesia" because it encourages visitors to recall a romanticized version of American history. Similar to what Michael Kammen terms the "heritage syndrome," rehabilitating battlefields to their historic condition seemingly presents the landscape in isolation by focusing only on the specific days relevant to a particular battle.[46]

The "grassing" of Gettysburg is both a careful covering over of perhaps unpleasant legacies in favor of a discrete national narrative, and, in the spirit of the traitor who "grasses" on criminal activity by revealing it to the authorities, it allows us to "read" the hidden history of this manufactured narrative on the very ground of the site. In *Leaves of Grass,* Walt Whitman writes,

> A child said *What is the grass?* fetching it to me with full hands;
> how could I answer the child? I do not know what it is any more than he.
> I guess it must be the flag of my disposition, out of hopeful green stuff
> woven.
> . . .
> Or I guess it is a uniform hieroglyphic,
> And it means, Sprouting alike in broad zones and narrow zones,
> Growing among black folks as among white,
> Kanuck, Tuckahoe, Congressman, Cuff, I give them the same, I receive
> them the same.
> And now it seems to me the beautiful uncut hair of graves.

. . .
O I perceive after all so many uttering tongues,
And I perceive they do not come from the roofs of mouths for nothing.

. . .
They are alive and well somewhere,
The smallest sprout shows there is really no death,
And if ever there was it led forward life, and does not wait at the end to
arrest it,
And ceas'd the moment life appear'd.[47]

The curated grasses at Gettysburg reflect Whitman's musings and connect the drive toward national identity with memory and the materiality of grass itself. Both deliberately planted and managed as well as continually pushing past boundaries and limits, the grass speaks in hieroglyphics that beckon us to understand something that is held just out of our line of sight. As both the "uncut hair of graves" and as "so many uttering tongues," the grass at Gettysburg is not only a reminder of death but is also the continuation of life. This dual presence points to the tension at the heart of grass management at Gettysburg: while the NPS may intend their management plan to project a particular kind of national identity, the grass itself may tell other stories.

Within the field of memory studies, scholars are considering memory in the context of the Anthropocene and how we might decenter the human and approach memory "ecologically (rather than merely socially)."[48] Doing so, Stef Creps contends, may require the field "to break with the persistent humanism that can be seen to prevent it from adequately addressing the vast spatio-temporal magnitudes of the Anthropocene."[49] Roseanne Kennedy similarly argues that scholars need to "exten[d] their objects to include the memory of nonhuman species."[50] While Kennedy is focused on nonhuman animals in her analysis, her call for thinking about how memory might work outside of the human can also call us to consider the ways in which plants might remember. Positioning ourselves in terms of "multidirectional eco-memory"[51] not only asks us to consider what a vegetal memory might be like but also draws our attention even more insistently to this truth: while management of grass and other plants can attempt to control the way that a landscape contains and transmits certain human memories and narratives, the life there will always extend beyond human control and suggest other memories and other histories.

In the poem that begins our chapter, Carl Sandburg's grass calls to us to "shovel [the bodies] under and let me work," reminding readers that the grass has a solemn role in meaning-making. So, where does meaning come from in a place like Gettysburg? Whose memories are highlighted, and why? What responsibility do we have to consider other voices, other memories, other histories? While our chapter cannot give voice to all of the stakeholders at Gettysburg, we can begin the process of considering how interdisciplinary approaches that take seriously the management of something as ubiquitous as grass can begin to unearth continually shifting and moving histories and identities of place.

Notes

1 "Plants," National Park Service.
2 "Plants," National Park Service.
3 Ladino, *Memorials Matter*, xii.
4 Ladino, *Memorials Matter*, xvii.
5 Cane, *Civilising Grass*, 3.
6 Alaimo, *Bodily Natures*, 9.
7 Nixon, *Slow Violence*, 30.
8 Alaimo, *Bodily Natures*, 9.
9 Murray, *On the Great Battlefield*, 10.
10 Murray, *On the Great Battlefield*.
11 Labbe, "Separate but Equal?"
12 Murray, *On the Great Battlefield*, 11.
13 Weeks, *Gettysburg*, 14.
14 Weeks, *Gettysburg*, 4.
15 Weeks, *Gettysburg*, 18.
16 Desjardins, *These Honored Dead*, 6.
17 Trowbridge, *The South*, 19.
18 Trowbridge, *The South*, 24.
19 Trowbridge, *The South*, 24.
20 Spielvogel, *Interpreting Sacred Ground*, 24.
21 In addition to the many battlefield tours on offer, Gettysburg hosts a dozen or more regular ghost tours, including "Blood on the Battlefield" (only for those ages beyond 16), "Echoes of War: All Ages Ghost Tour," "Haunted Orphanage: Ghostly Encounter Tour in Gettysburg," and more. ("Tripadvisor: Gettysburg Tours and Tickets").
22 "Foundation Document: Gettysburg National Military Park Pennsylvania," 3.
23 "Foundation Document: Gettysburg National Military Park Pennsylvania," 3.
24 Spielvogel, *Interpreting Sacred Ground*, 2.
25 "What We Do (U.S. National Park Service)."
26 Spielvogel, *Interpreting Sacred Ground*, 3.
27 Spielvogel, *Interpreting Sacred Ground*, 20.
28 "National Parks are Native Lands."
29 Spielvogel, *Interpreting Sacred Ground*, 152.
30 Spielvogel, *Interpreting Sacred Ground*, 106.
31 "Grasses."
32 "Grassland and Shrubland Birds," xi.
33 "Grassland and Shrubland Birds," 9.
34 Christians, Patton, and Law, *Fundamentals of Turfgrass Management*.
35 Ode, et al., "Indicators of Perceived Naturalness as Drivers of Landscape Preference"; Sharp, et al., "Factors Influencing Public Preferences for Invasive Alien Species Management."
36 Twedt, et al., "Designed Natural Spaces," 2.
37 "Turfgrass Maintenance in VA National Cemeteries," 9.
38 Bekken and Soldat, "Estimated Energy Use and Greenhouse Gas Emissions Associated with Golf Course Turfgrass Maintenance in the Northern USA"; Tidåker, et al., "Energy Use and Greenhouse Gas Emissions from Turf Management of Two Swedish Golf Courses."

39 Auwaerter, et al., "Cultural Landscape Report for Gettysburg National Cemetery: Gettysburg National Military Park," 129.
40 "Turfgrass Maintenance in VA National Cemeteries," E–1.
41 "Turfgrass Maintenance in VA National Cemeteries," 3.
42 Fetridge, Ascher, and Langellotto, "The Bee Fauna of Residential Gardens in a Suburb of New York City (Hymenoptera: Apoidea)"; Mallick, Rana, and Ghosh, "Role of Herbicides in the Decline of Butterfly Population and Diversity."
43 Spielvogel, *Interpreting Sacred Ground*, 112.
44 Spielvogel, *Interpreting Sacred Ground*, 112.
45 Melnick, "Climate Change and Landscape Preservation."
46 Murray, *On the Great Battlefield*, 183.
47 Whitman, *Leaves of Grass*, 29–30.
48 Creps, et al., "Memory Studies and the Anthropocene: A Roundtable," 500.
49 Creps, et al., "Memory Studies and the Anthropocene: A Roundtable," 500.
50 Creps, et al., "Memory Studies and the Anthropocene: A Roundtable," 506.
51 Creps, et al., "Memory Studies and the Anthropocene: A Roundtable," 506.

References

Alaimo, Stacy. *Bodily Natures: Science, Environment, and the Material Self.* Indiana University Press, 2010.

Auwaerter, John, Christopher Anderson, Catherine Ponte, Nathan Powers, Pamela Selby, and George Curry. *Cultural Landscape Report for Gettysburg National Cemetery: Gettysburg National Military Park.* National Park Service. Boston, MA: Olmsted Center for Landscape Preservation, 2017.

Bekken, Michael A. H., and Douglas J. Soldat. "Estimated Energy Use and Greenhouse Gas Emissions Associated with Golf Course Turfgrass Maintenance in the Northern USA." *International Turfgrass Society Research Journal* 14: 58–75. https://doi.org/10.1002/its2.61.

Cane, Jonathan. *Civilising Grass: The Art of the Lawn on the South African Highveld.* Wits University Press, 2021.

Christians, Nick, Aaron J. Patton, and Quincy D. Law. *Fundamentals of Turfgrass Management.* 5th ed. John Wiley & Sons, 2017.

Creps, Stef, Rick Crownshaw, Jennifer Wenzel, Rosanne Kennedy, Claire Colebrook, and Vin Nardizzi. "Memory Studies and the Anthropocene: A Roundtable." *Memory Studies* 11, no. 4 (2018): 498–515. https://doi-org.nmu.idm.oclc.org/10.1177/1750698017731068.

Department of Veterans Affairs. "Turfgrass Maintenance in VA National Cemeteries." *NCA Handbook 3420*, February 29, 2012. https://www.vendorportal.ecms.va.gov/FBODocumentServer/DocumentServer.aspx?DocumentId=2955388&FileName=VA786-16-R-0239-A00002008.pdf.

Desjardin, Thomas A. *These Honored Dead: How the Story of Gettysburg Shaped American Memory.* Da Capo, 2003.

Fetridge, Evelyn D., John S. Ascher, and Gail A. Langellotto. "The Bee Fauna of Residential Gardens in a Suburb of New York City (Hymenoptera: Apoidea)." *Annals of the Entomological Society of America* 101, no. 6 (2008): 1067–1077. https://doi.org/10.1603/0013-8746-101.6.1067.

Labbe, Savannah A. "Separate But Equal? Gettysburg's Lincoln Cemetery." *The Gettysburg Compiler: On the Front Lines of History*, Gettysburg College, March 12, 2018. https://gettysburgcompiler.org/2018/03/12/separate-but-equal-gettysburgs-lincoln-cemetery/.

Ladino, Jennifer K. *Memorials Matter: Emotion, Environment, and Public Memory at American Historical Sites.* University of Nevada Press, 2019.

Mallick, Buddhadev, Sukanta Rana, and Tuhin S. Ghosh. "Role of Herbicides in the Decline of Butterfly Population and Diversity." *Journal of Experimental Zoology*

Part A: Ecological and Integrative Physiology 339, no. 4 (2023): 346–356. https://doi.org/10.1002/jez.2688.
Melnick, Robert Z. "Climate Change and Landscape Preservation: A Twenty-First-Century Conundrum." *Journal of Preservation Technology* 40, no. 3–4 (2009): 35–42.
Murray, Jennifer M. *On a Great Battlefield: The Making, Management, and Memory of Gettysburg National Military Park, 1933–2023*. The University of Tennessee Press, 2023.
National Park Service. "Foundation Document: Gettysburg National Military Park Pennsylvania." August 2016. https://www.nps.gov/gett/learn/management/upload/GETT_FD_SP-508.pdf.
National Park Service. "Grasses." Accessed February 26, 2015. https://www.nps.gov/gett/learn/nature/grasses.htm.
National Park Service. "Plants." Accessed November 2, 2022. https://www.nps.gov/gett/learn/nature/plants.htm.
National Park Service. "What We Do (U.S. National Park Service)." Accessed February 27, 2025. https://www.nps.gov/aboutus/index.htm.
National Parks Conservation Association. "National Parks are Native Lands." Accessed July 21, 2025. https://www.npca.org/case-studies/national-parks-are-native-lands.
Nixon, Rob. *Slow Violence and the Environmentalism of the Poor*. Harvard University Press, 2013.
Ode, Åsa, Gary Fry, Mari S. Tveit, Pernette Messager, and David Miller. "Indicators of Perceived Naturalness as Drivers of Landscape Preference." *Journal of Environmental Management* 90 (2009): 375–383. https://doi.org/10.1016/j.jenvman.2007.10.013.
Peterjohn, Bruce G. "Grassland and Shrubland Birds of Gettysburg National Military Park and Eisenhower National Historic Site: Current Status and Management Recommendations." *United States Geological Survey*. December 2007. https://www.usgs.gov/publications/grassland-and-shrubland-birds-gettysburg-national-military-park-and-eisenhower.
Sandburg, Carl. "Grass." In *Cornhuskers*, vol. 126. Henry Holt and Company, 1918.
Sharp, Ryan L., Lincoln R. Larson, and Gary T. Green. "Factors Influencing Public Preferences for Invasive Alien Species Management." *Biological Conservation* 144 (2011): 2097–2104. https://doi.org/10.1016/j.biocon.2011.04.032.
Spielvogel, J. Christian. *Interpreting Sacred Ground: The Rhetoric of National Civil War Parks and Battlefields*. University Alabama Press, 2013.
Tidåker, Pernilla, Therese Wesström, and Thomas Kätterer. "Energy Use and Greenhouse Gas Emissions from Turf Management of Two Swedish Golf Courses." *Urban Forestry & Urban Greening* 21: 80–87. http://dx.doi.org/10.1016/j.ufug.2016.11.009.
Tripadvisor. "Tripadvisor: Gettysburg Tours and Tickets." Accessed March 23, 2025. https://www.tripadvisor.com/Attraction_Products-g60798-a_contentId.126414734838+14161075320-Gettysburg_Pennsylvania.html.
Trowbridge, John Townsend. *The South: A Tour of its Battlefields and Ruined Cities*. Hartford: L. Stebbins, 1866.
Twedt, Elyssa, Reuben M. Rainey, and Dennis R. Proffitt. "Designed Natural Spaces: Informal Gardens are Perceived to be More Restorative than Formal Gardens." *Frontiers in Psychology* 7, no. 88 (2016): 1–10. https://doi.org/10.3389/fpsyg.2016.00088.
Weeks, Jim. *Gettysburg: Memory, Market, and an American Shrine*. Princeton University Press, 2021.
Whitman, Walt. *Leaves of Grass: The First (1855) Edition*. New York: Penguin Books, 1986.

18 Exploring Traditional Ecological Knowledges in the Classroom

Indigenous Ecostudies

Cari M. Carpenter

I am the daughter of a biologist who taught me how to love the outdoors with scientific names in mind. My dad and I camped and hiked through the mountains of Colorado during my childhood, and every time I was quizzed on the plants we encountered, from the Kinnikinnick (which, my dad told me, Native Americans use) to the lupine (with its purple petals that are my father's favorite). Today, as an English professor invested in teaching nineteenth-century literature by Native American women, I have grown particularly interested in their ecoliterature. But I am also a non-native person, which makes this teaching all the more challenging and risky. As I continued to explore ecoliterature throughout my career, I began to learn more about TEK and began teaching it through Indigenous literature—the literature of Native Americans. Fortunately, this study of TEK coalesces well with literature, as anyone who teaches Native American Studies can attest. As Leslie Marmon Silko writes in the novel *Ceremony*, "[Stories] are all we have, you see./All we have to fight off/illness and death."[1] In this poem, the components of land management, knowledge about the world, are put together in a story. The ceremony is key to keeping the land—and thus the people that live upon it—safe.

From 2023 to 2024, I taught an honors Indigenous Ecostudies course that provides a way for professors to present resource management in a humanities context. It is with the wisdom of TEK—the Indigenous science of land management—that science becomes story. It emphasizes the land as part of us and, as such, key to our survival. Leonard et al. remind us that this belief system is inevitably multiple and diverse: "Indigenous understandings of the natural world and the importance of sustainable hunting, fishing, navigating, gathering, and cultivation practices are commonly referred to in academic literature as Traditional Knowledge, TEK, Indigenous Knowledge Systems, Native Science, or Ancestral Knowledge."[2] In effect, the knowledge a particular community develops is place-based and, therefore, entirely unique due to the land it springs from. The wisdom of TEK is thus never singular.

Throughout the course we discuss issues such as the following: how is TEK different from dominant approaches in environmental studies? What specific challenges do Indigenous communities face in maintaining and applying

DOI: 10.4324/9781003581185-23

their environmental knowledge systems? How can Indigenous perspectives on ecology and land management benefit both Indigenous communities and broader society? This course is designed for students from all majors who are interested in the intersecting fields of Ecocriticism, Native American Studies, and English. Specifically, it explores the history, current state, and future development of TEK—the current umbrella term for Indigenous approaches to the environment—such as living with the more-than-human world.

The course is divided into five units: Ethnology, Food Sustainability, Plant Ecology, Food Sovereignty, and Climate Change. As Indigenous literature like *Braiding Sweetgrass* or *Native Science* illustrates, TEK is grounded in several core beliefs: first, intelligence is everywhere, from animals to plants and even mountains and rocks, a belief known as "animism," or what Lakotas refer to as *tiospaye* [tee-yo'-shpaw-yay] ("all our relations"). TEK operates around the premise that time is circular rather than linear, as it is in the West. This way of thinking reminds me of a friend of mine who spent time on the Navajo Reservation and, one day, upon asking someone how she was doing, was told, "Not very good. The Long Walk is weighing on me."[3] I also think of a former student who, in reading the flashbacks in Silko's novel *Ceremony*, finally understood her own grandfather's memories of war in his journal. By writing his memories as if they were presently occurring, he was able to meet them in the current moment. Another characteristic of TEK is finding gratitude, or appreciation, for everything. As Robin Wall Kimmerer says in *Braiding Sweetgrass*, "We know that appreciation begets abundance. Why should it not be so for Mother Earth, who packs us a lunch every single day?"[4] TEK is also committed to a holistic, rather than a separating philosophy; it means opening one's mind to the unexpected. We explore chaos theory, an interdisciplinary area of science focused on underlying, unexpected patterns: deterministic laws of dynamic systems that are highly sensitive to initial conditions and as such difficult to control. Gregory Cajete writes,

> Rather than seeking to control natural reality, Native science focuses its attention upon subtle, inner natures wherein lie the rich textures and nuances of life. This is exactly what chaos theory shows us: small, insignificant things play major roles in the way a process unfolds.[5]

By beginning with a study of the basic premises of TEK, we can foster a love of natural resources in everyone. In a time of climate change, we might argue that such a love has never been more important. It is by learning TEK in the context of literary studies that students understand how essential land management is to themselves and are thus able to pursue its success as their own. The class I designed and taught requires attendance and participation, discussion posts each week, and a food and art display in which everyone is required to share an Indigenous food from a Native cookbook (Heid E. Erdrich's *Original Local* is the assigned version, but I also include readings from Sean Sherman's *The Sioux Chef's Indigenous Kitchen*). I am delighted with what the students produce for the art assignment: most did drawings, while one student played a piece on his tuba evoked by the assigned reading. The painting *Alaska* is an exceptional example of the artistry of many of my students.

The final assignment is either a traditional essay or a creative work that inspires several diverse projects: Indigenous menus, one of which is made from computer code; pottery with a thread of spider plant; a painting, a beautiful watercolor shot of fireweed in the Alaska wilderness (see Figure 18.1); and a short story. I am careful to make students aware of the problematics of appropriating Native skills or crafts into their projects. I do so by showing a miniature totem pole that a former student had made with a 3D printer and initiating a discussion about why this might be inappropriate for a person of another culture to make. We are thus able to discuss the totem pole's various meanings that those outside of a certain cultural context cannot fully understand or honor. I hope to make TEK more familiar with students by assigning a one-page response asking them to "describe TEK in your own words" or "discuss how the message of the poem relates to TEK." I design these assignments to provide students an opportunity to deal with the primary questions of the class in a low-stakes way.

As I begin my introduction to TEK early in the semester, I introduce a feeling with which students today might be familiar: the forgetting of and separation from the natural world that some urban residents might experience. It is my hope that if students begin the course with a feeling they can recognize,

Figure 18.1 "Alaska" painting by Calinda Evanovich [2025].

Source: Reproduced with permission of Calinda Evanovich.

they are more open to understanding and extending this feeling to different definitions of the land. I found this sense of rural estrangement in the first stanza of Cheryl Savageau's "Bones—A City Poem":

> [F]orget the great blue heron flying low
> over the marsh, its footprints
> still fresh in the sand
>
> forget that you are walking
> on the bones of your grandmothers.[6]

The entire poem becomes an opposition between the rural and the urban; however, in the end, readers realize the city has been the earth all along, and the city is in fact ourselves—"our grandmothers"—on which we walk. Ultimately, these words precisely articulate the theme that we are in fact intricately tied to nature, a message I hope resonates throughout the class.

Unit One: Ethnology

The class begins with an exploration of ethnology, the study of humans as cultural beings. This unit covers perhaps the most important part of TEK: our interdependence with one another and every living thing around us. Westerners often see the environment as separate from humans; however, practitioners of TEK do not. According to Martha Johnson, ethnology is "a body of knowledge built up by a group of people through generations of living in close contact with environment, and a system of self-management that governs resource use."[7] Gleb Raygorodetsky (in Gwich'in Elders) observes that "[s]piritual and ethical values have been woven into this knowledge, creating a system that has guided the people and helped them survive."[8] Aboriginal participants in the Circumpolar Aboriginal People and Co-Management Practice, a group of around 250 participants from Canada, the United States, Russia, Greenland, and England, were asked in 1996 to share their experience with wildlife management. They define ethnography as the following:

> Traditional knowledge is an accumulated body of knowledge that is rooted in the spiritual health, culture, and language of the people and handed down from generation to generation. It is based on intimate knowledge of the land, water, snow and ice, weather and wildlife, and the relationships between all aspects of the environment. It is the way people travel and hunt. It is a way of life and survival.[9]

Traditional knowledge is practical common sense, good reasoning, and logic built on experience. It is a system (a standard of conduct), setting out rules governing the use and respect of resources and an obligation to share. For example, it tells people that they do not have the right to hunt all animals of a species, as in wolf-kills, or other predator-control programs. The wisdom comes in ensuring that knowledge is employed in a way that benefits both people and the environment. Traditional knowledge is dynamic, yet stable, and is usually shared in stories, songs, dance, and myths.

In the ethnology unit, my students also discuss how in community, everyone is tied to a place. It is a place of interdependence and relational language, where leadership is earned through integrity, and one's sense of self is integrated with one's position in space. According to such a view, the natural world helps to create the community, and cosmology is a story of place, kinship, and environment. After introducing TEK and encouraging students to consider their relationship to the world, I turn to a poem about ethnology—Simon Ortiz's poem—"Culture and the Universe". In it, a person realizes a canyon wall as part of him (or her) self:

> We are wordless:
> I am in you.
>
> we are one self in the canyon.
> And the stone wall
> I lean upon spins me.[10]

TEK emerges quite clearly here; the individual is entirely interlaced with the canyon to such an extent that their connection is etched into the poem itself. This fusion is particularly apparent when the canyon speaks to the person as if both are the same self: it reads, "I am in you;" so there is even less of a distinction between the solid wall and the individual. Here, the stone wall becomes agentive, the spinning a transitional part of the dynamic convergence. Thus, the ethnology unit emphasizes how tied every individual is to the particular land he or she occupies.

Unit Two: Animals

Animals other than humans play the second act of this course, requiring four weeks of study and offering material that students, like the young woman who loved to photograph deer, are eager to cover. Though all humans are interconnected with their environments, Native American communities traditionally recognize less rigid distinctions between humans and nonhuman animals. They tell a wide spectrum of stories about animals: humans making treaties with them, marrying them, or even forming the basis of their existence. In some stories, nonhuman animals have souls, they help create the universe, and even mentor humans. As such, they are critical to a humanities-based land management course. Our very first reading in this unit, Cajete's *Native Science*, indicates the centrality of nonhuman animals to human life: Watakáme, a Huichol—a member of an Indigenous tribe of the Sierra Madre Occidental mountains of Mexico—lives "when animals take human form."[11] When humans began to be disrespectful to animals, Cajete writes, animals responded, limiting their communication to dreams and visions. Yet, nonhuman animals have remained, at least for Indigenous peoples, as earth mothers, shamans, talismans and amulets, rituals, and totems. Over 75% of Indigenous myth includes animals.[12] In class, I emphasize the reciprocal relationship between nonhuman animals and Native Americans by reading various stories and asking students to compare these relations with their own. Animism, a form of relation in which permission is asked of animals, requires strict ethics: fishers can only take larger fish, and hunters must take only what they carry and must avoid hunting certain kinds of animals like fawns.

After students have read the assigned stories on the different roles of nonhuman animals in Indigenous cultures, I assign them to groups with each focusing on a particular story. Each group is then presented with specific questions: How does this story fit the trickster story? Why is it so hard to believe in a story about nonhuman animals and humans being involved in a war council? Is it anthropocentric to believe this? Why do you think a tribe would believe a nonhuman animal could be an owner? Why do you think it's a deer that the character turns into? How differently would you feel about this animal if your tribal context was Choctaw? Considering different roles for animals in culture helps students see them in a new, expanded dimension.

On another day, I further expand these relations by making students aware of the Ojibwe belief in wolves as brothers to humans. I use "Activism and Indigenous Governance," an essay by Lint Carroll and Angelica Lawson, to analyze a situation in 2012 in which the states of Wisconsin and Minnesota ousted the gray wolf from the Endangered Species list and organized a wolf hunt without consulting resident Ojibwes, who strongly believe the wolf is their brother.[13] The Ojibwes developed a website to organize the Native activists against the hunt. After my students have read the essay, I ask the class to respond in writing to the following quote by Ojibwe traditionalist Edward Benton-Banai:

> What the Grandfather [Creator] said to them has come true. Both the Indian and the wolf have come to be alike and have experienced the same thing. Both of them mate for life. Both have a clan system and a tribe. Both have had their land taken from them. Both have been hunted for their hair. And both have been pushed very close to extinction. We can tell about our future as Indians if we look to the wolf.[14]

In the essay "Radiant Life," Linda Hogan also describes the close relationship between animals and humans, focusing on an injured eagle that once demanded her presence.[15] She describes how hunters must give thanks to animals when they are killed. Natural law, she argues, or agreements with animals, must exist. Hogan considers the ways animals have been disrespected: poached, slaughtered, or forced to survive in a contaminated environment. Hogan calls us to consider that environmental policy is better when developed over long, stable periods; when researchers have experienced problems and resolution; when there are consistent, self-generated aims; when it's corroborated by fishing science; when it's accompanied by experience. Above all, she calls for research-based policy derived from a balanced mix of fishers and hunters, scientists, and managers. Hogan's detailed, compassionate language about the mistreatment of animals leads students to lively philosophical questions about how they are—or should be—treated.

Unit Three: Plants

Plants, the focus of the third unit of the course, are also equal to humans according to many TEK perspectives. They serve a key role in peoples' lives as food, medical sources, and community. Describing the ritual of braiding sweetgrass, Kimmerer uses sacred words: "In our language it is called *wiingaaashk*, [wing-ashkw], the sweet-smelling hair of Mother Earth. Breathe it in and you start to remember things you didn't know you'd forgotten."[16] This braid, she says, is woven from particular strands: "Indigenous ways of knowing scientific knowledge, and the story of an Anishabekwe scientist trying to bring them together in service to what matters."[17] In this class, I use several readings and assignments to emphasize what it would mean to view plants not as banal objects, but rather, as living beings that are central to our lives. Joseph Bruchac, for example, calls us to listen to trees and notice—not objectify them. He refers to the "plant people," showing again the thin division between humans and nonhumans, as well as the story of the three sisters, with "the climbing beans, corn stalk sheltering and squash settling on the ground."[18] Describing the delicate balance between plants and people, he talks about how plants are aware; they are sent to heal the injured; good thoughts are required to find them; and certain people are said to come from trees. In the story "Thanks to the Trees," he refers to the medicinal plants from which people emerge. He tells of rituals focused on plants, like the Green Corn Ceremony and the Maple Sugar Gathering. He also relates a story of the Ojibwe people: the origins of manoomin, their beloved wild rice. In small groups, the students contrast the ways different plants are viewed by Indigenous and non-Indigenous people.

To get a sense of the local use of plants as medicine and food, the class takes a hike around the West Virginia University Core Arboretum, a 91-acre greenspace that includes a botanical garden and an old-growth forest; one tree that only recently died was once found to be alive in 1492. We are fortunate to be led by biology Professor and Arboretum Director Zach Fowler, noted herbologist, to see local plants that are medicinal or nutritional. I include in the syllabus books that detail these species, *Edible Wild Plants* by the West Virginia Department of Agriculture and *West Virginia Medicinal Plants, Trees, and Shrubs* by Bill Church. As Church details, Indigenous people modeled the use of certain local plants for a variety of medical problems:

> American Indians used the root tea for fevers, leaf poultice for hard to heal ulcers and scrofula. Iroquois women ate the raw leaves to prevent conception. Root poultice was used to draw out splinters and reduce swelling. Fresh and recently dried leaves and roots were considered emetic and expectorant.[19]

In order to bring these ideas to life, I have the students choose their favorite local medicinal and culinary plants and memorize their names. We spend particular time on the walk studying the bear corn plant, which has been used both by bears and by Native American women as a muscle relaxant: bears use it as a laxative in the spring after a long hibernation, while Indigenous women have used it to relieve menopausal symptoms.

Perhaps one of the easiest plants to identify and collect as food is the young "fiddleheads" of cinnamon and Bracken ferns. As I describe to the class, just rub off the fuzz and boil for 30 to 45 minutes in plain water. Then serve over toast with salt and butter as the dressing. Small pieces can also be added raw to salads.[20] Other local plants can be eaten, including ramps, pokeweed, marsh marigold, plantain, quickweed, cattails, fireweed, and many others. We eat several such plants in class, discuss our experience eating some, like leeks and dandelion, and write about the Indigenous stories of some flowers—all in an effort to make us more aware of the significance of plants to our survival.

Unit Four: Food Sovereignty

The fourth unit of the course is food, which is key to sovereignty: its proper production and distribution signal an independence of culture. Indeed, much research has shown that poor nutrition and diabetes are a mark of the dangers of colonialism.[21] I assign Heid E. Erdrich's cookbook, *Original Local*; I also encourage students to review Devon A. Mihesuah's *Recovering Our Ancestors' Gardens* and Sean Sherman's *The Sioux Chef's Indigenous Kitchen.* In cooking from these books, we discover the seeds of cultural resilience they contain. The Three Sisters (an Indigenous intercropping method of planting corn, beans, and squash together) are a perfect example of the values of interdependence and harmony. As Kimmerer describes,

> Of all the wise teachers who have come into my life, none are more eloquent than these, who wordlessly in leaf and vine embody the knowledge of relationship. Alone, a bean is just a vine, squash an oversized leaf. Only when standing together with corn does a whole emerge which transcends the individual. The gifts of each are more fully expressed when they are nurtured together than alone. In ripe ears and swelling fruit, they counsel us that all gifts are multiplied in relationship. This is how the world keeps going.[22]

Erdrich, in her collection *Original Local*, also notes hand-harvested manoomin and "good seeds" as valuable to Indigenous groups.[23]

To further emphasize the relationship between food and Native American cultures, I ask students to bring in a recipe made from one of the Indigenous cookbooks. The books are full of natural treats: consider, for instance, a dinner of raspberry mint bubbles, fiddleheads and quail heads in ramp nests on polenta, duck and bison cassoulet, and sour cream and craisin pie. To get everyone started, I bring in sunseed cornmeal cookies. We devote some discussion to the difficulty of finding all of the ingredients, like cattail pollen or arrowroot, in these dishes: a finding that reminds us of their unique origins.

Additionally, I assign literature from Camille Dungy's book, *Ghost Fishing*, which describes culinary items in delightful ways. There is, for example, Margaret Noodin's poem "Umpaowastewin" [ump-pah-wah-tay-win]:

> Ode'iminibaashkiminasiganke
> She makes strawberry jam
>
> ginagawinad wiishko'aanimad, waaseyaagami
> mixing sweet wind and shining water
>
> miinaa gipagaa nibwaakaa,
> with thick wisdom
>
> bigishkada'ad, dibaabiiginad
> pounding, measuring
>
> gakina gaa zha wenimangidwa
> everything we've cared for
>
> gakina gaa waniangidwa
> everything we've lost
>
> nagamowinan waa nagamoyaang
> the songs we have not yet sung
>
> miigwanag waa wawezhi'angidwa
> the feathers yet to decorate
>
> ezhi-zhoomiingweyaangoba
> and all the ways we've smiled
>
> mooshkine moodayaabikoong
> into jars filled to the brim
>
> ji-baakaakonid pii bakadeyaang.
> to be opened when we are thin.[24]

Here, strawberry jam, "sweet wind and shining water," that "thick wisdom," becomes "everything we've cared for," "everything we've lost." It is, ultimately, the savior when they are hungry. In order to emphasize the poem as an example of sovereignty—of survival—I play the author's reading of it, which is available on the Poetry Foundation website.[25]

Near the end of this unit on food, we analyze Kimmerer's description of pecans, which she argues offers a record of Native history:

> Nuts are like the pan fish of the forest, full of protein and especially fat—"poor man's meat," and they were poor. Today we eat them

daintily, shelled and toasted, but in the old times they'd boil them up in porridge. The fat floated to the top like a chicken soup and they skimmed it and stored it as nut butter.[26]

In order to reinforce the meaning of this story, I ask my students to find another culinary plant to trace in historic terms. One student, for example, tells a story that shows the historical significance of potatoes to his family. Again, food—another crucial part of TEK—tells a story about its followers.

Unit Five: Climate Change

I include an assortment of facts about contemporary damage to the land in the last portion of the course, emphasizing how Indigenous people understand the land. According to Gleb Raygorodetsky,

> The term "Land" is not restricted to the physical environment only. It has a much broader meaning, used by Indigenous people to refer to the physical, biological and spiritual environments fused together. The closest scientific equivalent of the "Land", taken without its spiritual component, is "ecosystem."[27]

Although land was once considered something key to our culture and way of life, it has since been abused. Actions like fracking, coal mining, and mountaintop removal are hard to imagine if one looks upon land as an extension of human life; for in endangering it, we endanger ourselves.[28]

Poems like Kimberly Blaeser's "Eloquence of Earth" describe the horrors of land destruction. For her, this destruction is utterly tied to language: "Like banned books words still burn on my tongue—reciprocity/sacred, preservation, earth, tradition, knowledge, protect."[29] She notes the history of language—"descendants of ajinaak and maang"—using words to record the traumatic destruction of land. In phrases like "flashing fins of poisoned fish" and "crumbled it in quick greedy fists," she reminds us that with its destruction comes forgetting (perhaps as Kimmerer speaks of forgetting the sweetgrass or Savageau writes of forgetting that you are walking on the bones of your grandmother): "how many will remember the eloquence of earth itself?" Words, she shows, are part of the trees themselves: she speaks of white birch, "bark marked/with sign, scrolls a history." And in Warren Cariou's "An Athabasca Story," we hear of one man's initial experience of a fracking operation:

> He had never seen anything like it. The only things moving on that vacant landscape were enormous yellow contraptions that clawed and bored and bit the dark earth and then hauled it away toward the big house. And the smell! It was worse than his most sulfurous farts, the kind he got when he ate moose guts and antlers. It was like being trapped in a bag with something dead.[30]

I read parts of the story in class and ask students to analyze it. I am delighted with their responses to the story: the fact, for example, that the fracking machine is described as a "house," which gives it a pleasant, familiar domestic feeling in contrast to their sense of the fracking industry.

I conclude the class with the complicated concept of climate change, the perfect storm that has marked the last centuries of the human occupation of earth. Sadly, scientists have shown how climate change is particularly dangerous to Indigenous communities, like those in the Arctic. According to Zoe Todd, scholars investigating climate change have simply not bothered to include Indigenous scientists like Sheila Watt-Cloutier in their studies. As Todd argues, the Inuits see what others consider "Gaia" as "Sila," the "breath that circulates out of everything."[31] That is, while Gaia is often understood as the planet—a fixed being—Sila is understood as a motion central to life, a *verb.*

Scholars must learn to include the critical work of Indigenous populations who are experiencing the ongoing effects of climate change. Successful mitigation of climate change will require not just paying attention to their specific place-based local knowledges but also thinking about why these knowledges are—in all of their variety—so useful. After all, focusing on a small area for a longer period of time requires a different kind of research than Western scientists typically observe. Many Western scientists, such as those on my guest speaker Wendy F. Todd's dissertation committee, for example, still tend to dismiss TEK. In contrast, those who more accurately understand TEK use that knowledge to limit or mitigate the earth's destruction. The Point au Chien of Louisiana, for instance, use maps to identify spaces most affected by climate change and build canals to stop storm surges, while the Odawas in Michigan conduct several ethnobotanical studies to see how climate change affects particular species.

I ask the students to write essays throughout the course to register their reflection on the material, and I hope that some will devote their final project to TEK approaches to climate change. Several have in the past, including one who studied Indigenous sovereignty as a solution to climate change. When I am confronted with the brutal reality in news about how the Gulf Stream will collapse anywhere between 2025 and 2090, unleashing a horde of lethal storms, I turn to the wisdom of Native Americans, who have already survived multiple apocalypses. As Julian Brave Noise Cat writes, "In a society built upon our graves, survival has become an act of resistance."[32]

Conclusion

This interdisciplinary class feels like exactly where I need to be. I've come to love sharing certain details with my students: chaos theory from a Native American point of view; what it means if humans aren't the only intelligent beings in the world; and learning to look at the universe with verbs instead of nouns. In the best moments of past classes, students have told me that Native literature has changed the way they see: the young Indigenous man from Alaska who cradled the picture of another Alaskan student at Carlisle Indian School, finding solace in the comparison; the woman who told me that Thomas King's "The Truth About Stories" gave her a new perspective of the relationship between parents and children.

I was fortunate to host a wonderful guest speaker in my class: Wendy F. Todd, an Alaska Native Haida from Hydaburg, Alaska. Her Haida name is K'ah Skaahluwaa, from the Xáadas Nation of the Sdast'as clan (Fish egg house). Dr. Todd is a geoscientist and oceanographer at the University of Duluth. In 2019, she was named the American Indian Science and Engineering Society Professor of the Year for her interdisciplinary research. In my class, she offered a wealth of information about her research and her personal struggles in her graduate work: at her dissertation defense, she was informed that the faculty would not read her article on TEK, despite its importance to the overall project.

On the last day of class, I asked the students to write about how the course had changed their lives. One student responded,

> I feel my life has already changed. I look at my surroundings with more compassion, I have more interest in Native American culture, and I have learned about many current issues I had no idea about before. I will continue to use the things I have learned in this class in order to make me a better person.

Another student wrote,

> I definitely will change aspects of my life outside of class after taking this class. My knowledge of TEK has taught me to appreciate the relationships that occur in nature, and all the beauty in those relationships as well. I'm hopeful that I will continue to learn TEK more in the future and deepen my understanding of not only what nature offers to us, but what I can offer to it.

Finally, one remarked,

> This class has gotten me to go outside and connect with nature. I have realized that it makes me calm and collected, even if it's a short moment of time. A brisk walk or doing homework on the grass makes a Huge difference. I have also learned how to better understand TEK beliefs. With this information, I have formed a new view of the world, outdoors, and other cultures around me.

As I said in the beginning, this teaching is for me a process of sharing such stories. Whether you are the biology student who learns to join your experience with narrative, the English major who joins stories with scientific names, or the climate change activist who wants to view the melting glaciers with a different gaze, what you learn in this class is largely up to you. It's like what Robin Wall Kimmerer says of her project:

> It is an intertwining of science, spirit, story—old stories and new ones that can be medicine for our broken relationship with earth, a pharmacopoeia of healing stories that allow us to imagine a different relationship, in which people and land are good medicine for each other.[33]

I approach this class in Daniel Heath Justice's terms, "imagining otherwise"[34] my relationship to the earth as I pursue a love for TEK. I do so not without trial, as I acknowledge that my relationship to Native Americans is not without some unfamiliarity. Yet, this love is sufficient to foster an equitable ecological view that unsettles colonialism, which I ultimately hope to share with my students.

Spring 2024 HONORS 207

Indigenous Ecostudies

This course, Indigenous Ecostudies, is designed for students from all majors who are interested in the intersecting fields of Ecocriticism, Native American Studies, and English. Specifically, it will explore the history, current state and future development of TEK—the current umbrella term for approaches to the environment. The course will include both technical information (TEK) and supporting literature. Throughout, we will discuss issues such as how is TEK different from dominant methods of environmental studies? What particular challenges do Indigenous people have in these areas? How can their ideas benefit their own and wider communities? I will implement a variety of creative resources in order to bring Native views directly into the classroom, including Skypes or class visits with Indigenous scientists.

Course Goals

By the end of the quarter, I hope you will be able to:

- Identify specific ways that TEK differs from Western science.
- Develop creative responses to the environment and analyze others' written responses.
- Evaluate methods of TEK involving animals, plants, food, land management, and climate change.

Requirements

1. *Attendance and Participation (20% of final grade).*
2. *Food and Art Contribution (10% of final grade)*
3. *Comments on Discussion Board (10% of final grade)*
4. *First Exam (15% of final grade)*

5. *Second Exam (15% of final grade)*
6. *Final Project (30% of final grade)*

Schedule

T Jan. 9: *Introduction and Ethnology*
R Jan. 11: Cajete, "Introduction" (10–45)
T Jan. 16: McGregor, "Traditional Ecological Knowledge"
R Jan. 18: Menzies and Butler, "Introduction"; Berkes, "Introduction"; Savageau, "Bones"
T Jan. 23: Kimmerer, "The Grammar of Animacy"
R Jan. 25: Cajete, "The Ecology of Native Community"; Ortiz, "Culture and the Universe"

Animals

T Jan. 30: Cajete, "Animals in Native Myth and Reality"
R Feb.1: King, "The Truth About Stories"
T Feb. 6: Hogan, "Radiant Life"
R Feb. 8: *Exam One*
T Feb. 13: "The Marriage to a Bear"; "Strange Births"; "Water-Panther Husband" R Feb. 15: Berkes, "Cree World 'From the Inside'"
T Feb. 20: Guest Speaker
R Feb. 22: "Coyote Challenges the Snake"; "The War Between the Animals and Birds"; "The Owner of the Animals"; "The Man Who Became a Deer"

Plants

T Feb. 27: guest speaker
R Feb. 29: *Edible Wild Plants* (vooks021/01/Edible-Wild-Plants.pdf); *West Virginia Medicinal Plants, Trees, and Shrubs* (https://www.angelfire.com/bc/bchurch/Herb_book.PDF); Carroll and Lawson, "New Media and Environmental Governance"
T Mar. 5: Bruchac, "Listening to the Plants"; "Bitterroot"; "Thanks to the Trees"; "The Woman Who Lives in the Earth"; "Noman-shua"
R Mar. 7: Kimmerer, "A Gift of Strawberries"; Cajete, "Plants, Food, Medicine, and Gardening"
T Mar. 19: Kimmerer, "Asters and Goldenrod," "The Three Sisters"; "The Origin of Maize"
R Mar. 21: Mihesuah, "Introduction"; Noodin, "Umpaowastewin"

Food Sovereignty

T Mar. 26: Stanciu, "Food Sovereignty"; Sherman, "Introduction"
R Mar. 28: Kimmerer, "The Council of Pecans"
T Apr. 2: *Exam Two*

Climate Change

R Apr. 4: Cariou, "Athabasca Story"; Berkes, "Climate Change and Indigenous Ways of Knowing"
T Apr. 9: Wildcat, *Red Alert*; Stoffle, "Great Lakes Ecology"
R Apr. 11: Wildflower walk Cut
T Apr. 16: Dimaline, *Marrow Thieves* ("Frenchie's Coming-To Story"-"Story Part 1")
R Apr. 18: Dimaline, *Marrow Thieves* ("Magic Words"-"The Four Winds")
T Apr. 23: Dimaline, *Marrow Thieves* (to end); Blaeser, "The Eloquence of Earth"
R Apr. 25: Todd, "An Indigenous Feminist's Take on the Ontological Turn"

Notes

1 Silko, *Ceremony*, 2. Kristen Brown analyzes the ecorelational literacy—the loving model of land that unsettles colonialism—of *Ceremony* in the chapter, "Back to Belonging': Sound, *Ceremony*, and Resonance in a Re-membered Commons" in this volume.
2 Leonard, et al., "Water Back," 347–428.
3 Mitchell, interview by Cari M. Carpenter, January 2011.
4 Kimmerer, *Braiding Sweetgrass*, 116.
5 Cajete, *Native Science*, 17.
6 Savageau, "Bones: A City Poem," 67.
7 Johnson, *Lore*, 4.
8 Quoted in McGregor, "Traditional Ecological Knowledge," 78.
9 Quoted in McGregor, "Traditional Ecological Knowledge," 78.
10 Ortiz, "Culture and the Universe," 104–105.
11 Cajete, *Native Science*, 11.
12 Cajete, *Native Science*, 166.
13 Carroll and Lawson, "New Media and Environmental Governance," 119–135.
14 Carroll and Lawson, "New Media and Environmental Governance," 130–131.
15 Hogan, "You Call It Animism, We Call It Tradition." When I contacted Logan via email in February of 2024, Hogan did not think this chapter was published in 2018. I do not recall its web address; I call it "Radiant Life" as that's how it appeared online.
16 Kimmerer, *Braiding Sweetgrass*, ix.
17 Kimmerer, *Braiding Sweetgrass*, ix.
18 Bruchac, *Native Plant Stories*, xx.
19 Church, *West Virginia Medicinal Plants, Trees, and Shrubs*, 6.
20 Gillepsie, *Edible Wild Plants*, 5.
21 Mihesuah, *Recovering Our Ancestors' Gardens*.
22 Kimmerer, *Braiding Sweetgrass*, 140.
23 Erdrich, *Original Local*, 246.
24 Noodin, "Umpaowasetewin," 10–11. Permission to reprint from Wayne State University Press.
25 Noodin, "Umpaowasetewin," https://www.poetryfoundation.org/poetrymagazine/poems/58648/umpaowastewin.
26 Kimmerer, *Braiding Sweetgrass*, 13.
27 McGregor, "Traditional Ecological Knowledge," 79.
28 These impressions of the relationship between land destruction and the endangerment of people are related to Jessica Cory's work in this volume's chapter,

"Violence against the Land is Violence against the People: Land Management Tactics in Linda Hogan's *Solar Storms*."

29 The lines from this poem come from Blaeser, "Eloquence of Earth,"95–96.
30 Cariou, "An Athabasca Story," 70–75.
31 Todd recalls a 2013 conference in which the keynote speaker, the famous philosopher Bruno Latour, gave a keynote address on climate change that lacked any mention of the contributions of Indigenous scientists to the study. See Todd, "An Indigenous Take on the Ontological Turn," 4–22.
32 Brave Noise Cat, "Indian Country: A Postapocalyptic People," 30–33.
33 Kimmerer, *Braiding Sweetgrass*, 12.
34 Heath Justice, *Why Indigenous Literatures Matter*, 156.

Bibliography

Berkes, Fikret. *Sacred Ecology*. Routledge, 2012.
Blaeser, Kimberly. "Eloquence of Earth." In *Copper Yearning*. Holy Cow! Press, 2019.
Bruchac, Joseph. *Native Plant Stories*. Fulcrum Books, 1995.
Cajete, Gregory. *Native Science: Natural Laws of Interdependence*. Clear Light Publishers, 2000.
Cariou, Warren. "An Athabasca Story." *Lake: A Journal of Arts and Environment* 7 (2012): 70–75.
Carroll, Lint, and Angelica Lawson. "New Media, Activism, and Indigenous Environmental Governance." In *Ecocriticism and Indigenous Studies: Conversations from Earth to Cosmos*, edited by Salma Monani and Joni Adamson. Routledge, 2015.
Church, Bill. *West Virginia Medicinal Plants, Trees, and Shrubs: A Field Guide*, x-3. Lulu Books, 2004. https://www.angelfire.com/bc/bchurch/Herb_bookPDF.
Dimaline, Cherie. *The Marrow Thieves*. Cormorant Books, 2017.
Erdrich, Heid E. *Original Local: Indigenous Foods, Stories, and Recipes*. Minnesota Historical Society, 2013.
Gillepsie, William H. *Edible Wild Plants*. West Virginia Department of Agriculture, n.d. https://agriculture.wv.gov/wp-content/uploads/Edible-Wild-Plants.pdf.
Gwich'in Elders. *Nanh'Kak Geenjit Gwich'in Ginjik: Gwich'in Words about the Land*. Gwich'in Renewable Resource Board, 1997.
Hogan, Linda. "You Call it Animism, We Call it Tradition," 2018 (Unpublished).
Johnson, Martha. *Lore: Capturing Traditional Environmental Knowledge*. Dene Cultural Institute and the International Development Research Centre, 1992.
Kimmerer, Robin Wall. *Braiding Sweetgrass: Indigenous Wisdom, Scientific Knowledge and the Teaching of Plants*. Milkwood Editions, 2015.
King, Thomas. *The Truth About Stories: A Native Narrative*. House of Anansi Press, 2003.
Leonard, Kathey, Dominique David-Chavez, Deondre Smiles, et al. "Water Back: A Review Centering Rematriation and Indigenous Water Research Sovereignty." *Annual Water Alternatives Review* 16, no. 22 (2023): 374–428.
Lint, Carroll, and Angelica Lawson. "New Media and Environmental Governance." In *Ecocriticism and Indigenous Studies: Conversations from Earth to Cosmos*, edited by Salma Monani and Joni Adamson. Routledge, 2015.
McGregor, Deborah. "Traditional Ecological Knowledge and Sustainable Development: Towards Coexistence." In *The Way of Development: Indigenous Peoples, Life Projects and Globalization*, edited by Mario Blaser, Glenn McRae, and Harvey Feit. St. Martin's Press, 2004.
Menzies, Charles R. "Understanding Ecological Knowledge." In *Traditional Ecological Knowledge and Natural Resource Management*, edited by Charles H. Menzies. University of Nebraska Press, 2006.

Mihesuah, Devon A. *Recovering Our Ancestors' Gardens: Indigenous Recipes and Guide to Diet and Fitness*. University of Nebraska, 2020.
Noisecat, Julian Brave. "Indian Country: A Postapocalyptic People." *The Nation* 15, no. 22 (June 2020): 30–33.
Noodin, Margaret. "Umpaowastewin." In *What the Chickadee Knows*. Wayne State University Press, 2020.
Ortiz, Simon J. "Culture and the Universe." In *Out There Somewhere*. University of Arizona Press, 2004.
Savageau, Cheryl. "Bones: A City Poem." In *Dirt Red Home*. Curbstone Press, 1995.
Sherman, Sean, et al. *The Sioux Chef's Indigenous Kitchen*. University of Minnesota Press, 2017.
Silko, Leslie Marmon. *Ceremony*. Penguin Books, 1977.
Stanciu, Cristina. "The Recovery of the People Is Tied to the Recovery of Food: Food Sovereignty and Winona LaDuke's *Last Standing Woman*." *East-West Cultural Passage* 19, no. 2 (December 2019): 121–139.
Todd, Zoe. "An Indigenous Feminist's Take on the Ontological Turn." *Journal of Historical Sociology* 29, no. 1 (March 2016): 4–22.
Van Vlack, Kathleen A. "A Native Knowledge of Great Lakes Ecology: Climate Changes to Odawa Lands." *Frontiers in Climate* (March 5, 2023): 1–15.

19 "For Sheer Joy in Wild Terrain"

Rock Climbing Literature and Public Lands

Sarah Jane Kerwin

In the fall of 2022, climber and businessman Yvon Chouinard announced that he was giving away his multibillion-dollar outdoor apparel and gear company. "Earth is now our only shareholder," a banner at the top of the Patagonia homepage reads (emphasis original).[1] Patagonia has made similar environmentalist moves in the past, including suing the Trump administration over its decision to reduce the size of the Bears Ears and Grand Staircase-Escalante National Monuments in 2017, but donating all future profits to fighting the climate crisis is by far Chouinard's most direct assertion that he is committed to planetary health over personal gain. "Instead of 'going public,'" he declares, "you could say we're 'going purpose.'"[2]

Cringe-worthy language aside, the structure of Chouinard's approach to "going purpose" is admirable. As more and more companies market their products through greenwashing, Patagonia is legitimately putting its money where its mouth is. The action will not result in tax breaks for the Chouinard family, and keeping the company's profits in a 501(c)(4) organization means the money can be used for lobbying and political donations.[3] Patagonia is trying its best to fight the good fight.

But Patagonia's relationship to the environment is not as straightforward as Chouinard might like it to be. As ecocritic April Anson astutely points out, Patagonia's 2017 ad campaign proclaiming "The President Stole Your Land" relies upon settler-colonial logics to rally outdoor enthusiasts to the cause of protecting public lands. Anson contends that the "your" in the ad implicitly appeals to "white settler subjects," inviting us to view the national monument shrinkage "as an *initial* moment" of theft while ignoring the "past and ongoing" theft of Native lands.[4] Meanwhile, Chouinard is one of the best known rock climbers in history: his name is associated with the so-called "Golden Age" of Yosemite climbing in the 1950s and 1960s, the "Clean Climbing" movement of the 1970s, and the development and production of cutting-edge climbing gear.[5] Rock climbing has long been perceived as going hand-in-hand with environmentalist efforts that privilege the preservation of mountain wildernesses specifically.[6] The influence of Romanticism on early American preservationists like John Muir resulted in a heightened celebration of sublime landscapes, those dramatic enough to induce wonder

DOI: 10.4324/9781003581185-24

bordering on terror. The idea that fear-inducing encounters can lead to a kind of enlightenment permeated early climbing culture and persists to this day among outdoor enthusiasts, although few would describe their experiences as overtly religious. Instead, contemporary climbers, skiers, and the like approximate a similar feeling in the use of words like "stoke," which Ethan Linck defines as "feeling high on the pleasure of movement, the thrill of exposure." As Linck asserts in his *High Country News* article "Your Stoke Won't Save Us," outdoor recreation culture, with its "fixation on rock and ice," perpetuates an infatuation with the alpine wilderness, to the detriment of more nuanced conceptions of what nature and environmentalism can and should mean.[7]

Linck and many others have suggested that we need to spend more time both existing in and thinking about the environments of our own backyards, a call for letting go of the idea that the ecosystems most worth protecting are those that sweep us off our feet.[8] In the United States, which has long deemed itself *nature's nation*, we need to dedicate more energy toward caring for and protecting the nature that exists in places that are perceived as ordinary, accessible, and perhaps even flat.[9] However, I believe mountain enthusiasm can continue to play an important role in contemporary environmental efforts, in large part, because rock climbing's popularity has grown significantly in recent decades.[10] Although much of this growth has occurred in climbing gyms, most gyms invite even the newest indoor climbers to imagine themselves as part of an outdoorsy community. Rock climbing has also benefited from a substantial online presence, and both professional and amateur climbers use digital media and social media as platforms to advocate for increased access and to raise awareness about environmental issues.

For example, Cat Runner, an Asian American transgender athlete who won the 2023 HBO climbing competition series *The Climb*, uses his social media presence to promote the Queer Climbers Network, which he founded in 2022.[11] In August 2021, Kathy Karlo interviewed Runner on her podcast *For the Love of Climbing*, which she describes as "a podcast about choosing vulnerability and talking about our pain—and how we are all really just shining examples of this messy human existence." In the spirit of vulnerability, Runner reflects on moments when he has wondered, "Am I here because of my ability to do something, or am I here because I am the diversity that they wanted?" Although he emphasizes later in the episode that he is trying to "step away mentally from grades" and focus on "the movement" of climbing, Runner's victory in *The Climb* serves as an important reminder that he was selected for the TV show because of his "ability to do something."[12] Along with promoting equity of access, the Queer Climbers Network and other groups working to diversify outdoor recreation are telling new stories about what climbers and climbing can look like.

Higher profile climbers are also beginning to engage with hard questions about climbing culture. Alex Honnold and Fitz Cahall's popular podcast *Climbing Gold* tackles the issue of rock climbing on Native lands in

"Chapter Nine: Untelling the Story." The episode features a number of Indigenous climbers, including Natives Outdoors founder Len Necefer (Diné), and focuses primarily on the Navajo Reservation, where all climbing is illegal. Cahall also reminds listeners that "almost 60% of climbing [in the U.S.] exists on public lands" and calls attention to the fact that although these lands are not under Native control, many popular crags "have deep cultural significance to tribes," including Bear Lodge in Wyoming.[13] In 1995, the NPS implemented a voluntary ban on climbing Bear Lodge during the month of June, to moderate success.[14] As "Untelling the Story" makes clear, climbers have had a historically bad track record when it comes to recognizing and respecting Indigenous land rights, but these limitations are not baked into the physical act of climbing and can be changed for the better. The heightened environmental perception climbing facilitates, on the other hand, is inherent to the pursuit, which is what makes climbing culture such a compelling subject of ecocritical study. Legitimate inclusion and social justice efforts will only expand the possibilities for environmental advocacy within the sphere of climbing.

As experts in a variety of fields—including wildlife and plant biology, forestry, and the outdoor recreation industry—work to figure out the best practices for sustainably managing the natural resources of alpine and subalpine public lands in North America, it is the job of ecocritics to engage with what these spaces have meant for humans in the past and will continue to mean moving forward. Climbing literature, broadly defined, should be included in this ecocritical conversation.

This chapter focuses on the writings of rock climbers Doug Robinson and Joe Kelsey, both of whom weave mindfulness, environmentalism, and an appreciation for the written word into their descriptions of what Robinson calls "moving over stone."[15] I begin with Robinson's *A Night on the Ground, A Day in the Open*, which was published in 1996 and contains essays spanning four decades of climbing and skiing in the Sierras, then turn to Joe Kelsey's *A Place in Which to Search: Summers in the Wind Rivers* (2016). Whereas I read *A Night on the Ground* with an ecocritical lens to make a case for its environmentalist potential, I argue that *A Place in Which to Search* is an important work of environmental literature deserving of a far wider readership than the backpackers who have a habit of scanning niche interest sections of small Western bookstores. Robinson and Kelsey invite readers to consider the ways in which outdoor physical pursuits can lead to *active care* for the public lands in North America where so many people go to climb.

When I spoke with Robinson about his more recent rock scrambling endeavors as an older man, he emphasized this point:

> I might not even carry water, [but] I do carry my notebook—my best writing has been spontaneously jotted down in the notebook. . . . It's a practice to me and I've noticed it's at its best when I'm moving in the mountains.[16]

I propose rethinking our collective understanding of land management so as to include not only actions with direct objectives in the field—like invasive species removal and controlled burns—but also more indirect practices such as climbing, hiking, reading, and writing.

From "Clean Climbing" to Mountaineering Redefined

Doug Robinson is perhaps best known for his role in helping to spark what has become known as the "Clean Climbing revolution" with his article "The Whole Natural Art of Protection." Originally published in the 1972 *Chouinard Equipment Catalog* and republished in *A Night on the Ground*, Robinson's article/manifesto proclaims that climbers should stop hammering and bolting gear into the rock and instead use only temporary protective gear. This movement was inspired at least in part by environmental concerns, a commitment to finding ways, in Robinson's words, to "protect the vertical wilderness from human erosion." "Clean is climbing the rock without changing it," Robinson asserts, "a step closer to organic climbing for the natural man."[17] Such a purist mindset is bound up with the idea that there are right and wrong ways to encounter nature and, more importantly, that men who have demonstrated some kind of socially and culturally rewarded prowess in their chosen pursuit are the rightful arbiters of what those right and wrong forms entail. It should be noted here that all forms of climbing have an environmental impact on the rock itself as well as on animals, plants, and soil. Studies have shown that climbing can cause disturbances to cliff-nesting birds, damage lichen and bryophytes on cliff faces, and reduce vegetation at the bases of cliffs, leading to concerns about reduced biodiversity.[18] Because truly clean climbing is unattainable, both climbers and the environment will be better served if we follow Karlo's lead and grapple with the human messiness of climbing, past and present. Despite its shortcomings, there is still much to be admired in the environmental ethos underlying Robinson's manifesto, an ethos that runs throughout *A Night on the Ground*.

Robinson believes climbing, skiing, and other outdoor activities are "channels for applying yearning to landscape": they "intensify contact with the mountains."[19] The odd phrasing here—"applying yearning *to*" what would typically be yearned *for*—indicates that the object of Robinson's yearning is not the mountains themselves, but rather the "intensif[ied] contact" with the mountains that moving within them brings (emphasis added). Desire is certainly present in this yearning, but it is not possessive. Rather, Robinson desires to follow the sound advice he received from his mother as a child: "Never lose your contact with the earth." Robinson recalls her saying this while they were seated on a "granite slab" beside Tenaya Lake in Yosemite, a reminder to readers that such contact might come about in myriad ways: on the granite face of Yosemite's walls, yes, but also in the stillness of sitting by a lake.[20]

Robinson builds this more expansive perspective on what it means to "intensify contact with the mountains" into the version of "The Whole Natural Art of Protection" included in *A Night on the Ground* by adding a single sentence to the end: "Right from the start, clean climbing demands increased awareness of the rock environment."[21] For Robinson, developing a sense of "responsibility" for protecting the "rock environment" is inherently tied to a state of heightened attentiveness brought about by climbing. Here, he stresses the "increased awareness" that comes from climbing without fixed protection, but other essays in *A Night on the Ground* explore alternative pathways to becoming more attuned to one's environment.[22]

In the essay "The Climber as Visionary," Robinson declares, "To climb with intense concentration is to shut out the world, which, when it reappears, will be as a fresh experience, strange and wonderful in its newness."[23] Climbers alternate between periods of active focus—the climb itself—and more passive periods while they belay their partner; Robinson points out that while climbers belay, they have time to notice and reflect on the surroundings they were so physically imbricated in minutes prior. In "Why Climbing Matters," Jeffrey Mathes McCarthy argues that climbing "can reveal that humans and nature intermingle, overlap, connect, instead of exist side by side as active subject and passive object."[24] McCarthy, whose article briefly engages with "The Climber As Visionary," highlights "the body" as "the active vehicle of recognition, the means by which the world becomes intertwined with the human."[25] As Robinson moves his body up a wall, his ability to read the tiniest details of its contours, to find an available ledge to balance a toe or a pocket where he can fit two fingers inside, is what will determine whether he reaches the top or falls. Once he has safely reached the top of the pitch, Robinson asserts, his mind will be able to see not only what his body has already recognized but also beyond it: this is when the climber is able to notice "what is more deeply interfused," a way of seeing that Robinson describes as a kind of ecology. "It is an intuitive rather than a scientific ecology," he explains, "starting not from generalizations for trees, rocks, air, but rather from that tree with the goiter part way up the trunk."[26]

In the chapter "Grassing Gettysburg: Management, Memory, and Meaning," Michael R. Barnes and Amy T. Hamilton highlight "reading both management policies and land as texts." Although Robinson does not describe the rock as a text, the alternating actions of climbing and belaying facilitate a *rock literacy* of sorts. The tactile nature of rock literacy is what makes it unique: reading the rock as a climber requires physical touch. "The Climber as Visionary" mediates the rock climbing/rock reading experience for readers, while also indicating the gap—or perhaps the crack—between reading this essay and reading granite for oneself.

While the sense of interconnectedness McCarthy calls attention to is crucial for understanding Robinson's ecological perspective, it is imperative to expand beyond climbing narrowly defined—scaling a rock face—in order to open up a greater array of possibilities for ways to "intermingle, overlap, and connect" with mountain environments. To do so, I trade the term "climbing" for "mountaineering," drawing inspiration from a poem by Robinson:

> Mountaineering just means glad to be here
> No climbing is required.
> Simply from being in the mountains
> It will arise spontaneously of itself,
> For sheer joy in wild terrain.[27]

This short poem expresses the idea at the heart of *A Night on the Ground*, which is that spending days in the open is the core purpose of all Robinson's physical pursuits. Here, mountaineering does not require absurd grip strength or massive gear racks; it simply depends upon "being in the mountains." The climber who rushes every step of the way to accomplish a difficult route may in fact be less of a mountaineer than someone who picks their way carefully across a trail to find a good vantage point for perching, breathing, looking, and feeling glad.[28]

Despite the hours he spends leaning against "granite backrests" writing in his notebook, Robinson nevertheless maintains that the "oceanic feeling of the supreme sufficiency of the present" is best achieved through concentrated physical effort.[29] He does not, however, construct a hierarchy in which the most elite climbers have the best access to such embodied present-ness. Instead, Robinson underscores the importance of developing an intentional practice of moving through the mountains, which can take a range of forms depending on one's abilities. In an essay called "Running Talus," Robinson makes a case for "rambling along rocky trails" as a way of merging bodily rhythm with the environment without the need for ropes or specialized skills or even a clearly achievable end goal.[30] Participating in "a personal and noncompetitive kind of mountaineering," the talus runner seeks a balance-in-motion determined by the spacing of rocks along a pathway the runner chooses as they go. This practice can be an early training step on the way to more advanced mountaineering, but it can also be a practice in the sense that it is a pursuit in its own right: talus running, as much as climbing, "accelerates attention closer to the present moment, where clinging to a stray thought in the instant between perception and motion can lose the climber his finely tuned edge and send him crashing. Staying on that edge, however, breeds mindfulness."[31] Such mindfulness, rather than stoke or a fixation on purity, is what mountaineers like Robinson bring to the environmentalist table.

"Something to Wonder about": Searching in the Winds

Early in *A Place in Which to Search*, Joe Kelsey notes, "We like to imagine climbing as a direct interaction between climber and rock, but most climbers I know have had this interaction profoundly influenced by the printed word."[32] He goes on to articulate the importance of "climbing literature," such as accounts published in the *American Alpine Journal* (*AAJ*) in the 1960s, which established both classic routes and a "mythology" around rock climbing.[33] While Kelsey acknowledges the significance *AAJ* accounts held for him as a young climber, he soon broadens the kinds of texts that helped him "understand the life [he] had chosen," ranging from histories of the US West to canonical works of nature writing: Edward Abbey's *Desert Solitaire,* Annie Dillard's *Pilgrim at Tinker Creek*, and Norman Maclean's *A River Runs Through It.*[34] "These three books," Kelsey states, "were the ones that gave me the most meaningful perspective on a Wind River climber's life."[35] For Kelsey, rock climbing, literature, and environmental awareness are inseparable from one another.

A Place in Which to Search is a difficult book to pin down. It documents nearly five decades of summers in Wyoming's Wind River Mountains, but does so without a clear chronology. Instead, chapters are arranged thematically, an organizing scheme which is more easily legible in some instances than others. For example, "The Cabin" centers almost entirely around the small off-grid cabin that has been Kelsey's summer home base since the mid-70s, whereas "Mushroom Soup" takes its title from a single anecdote and then opens out onto a range of meditations around "the why of climbing."[36] The subject of "Bad Weather," meanwhile, is not nearly as simple as its title implies. Although it could be called a memoir, *A Place in Which to Search* is also a collection of field notes, a natural history, a work of philosophy, and a love letter: to the Winds themselves, to the community that comes together there, and to the "Wind River state of mind."[37] It is also an important contribution to the tradition of environmental writing about the US West.

The final sentence in the Preface sets the tone for the rest of the book: "That a chunk of rock thrust up from the earth's crust is named for flowing water that is named for moving air seems appropriate."[38] After grounding us in the mountain range—"a chunk of rock thrust up from the earth's crust"—Kelsey shifts into the realm of words, pointing out that these mountains take their name from the river to the northeast, which in turn takes its name from an even more ephemeral entity: "moving air." Such ephemerality does not negate the serious force of Wyoming wind, as Kelsey is well aware. By pointing out the import, both linguistically and materially, of something so difficult to pin down, Kelsey suggests that the Winds are similarly impossible to grasp. Kelsey does not attempt to contain the Winds within the pages of *A Place in Which to Search*, thereby refusing to stake any sort of claim on this mountain range. Instead, he strives to bring the reader into the Winds, at least for a little while.

Like Robinson, Kelsey embraces the idea that because rock climbing demands entirely focused attention, climbers develop a heightened awareness of their surroundings, particularly when paired with "a belayer's opportunity for reflection."[39] Such awareness does not automatically translate to anything beyond the goal of moving up a rock wall or mountain, but Kelsey believes the "intense involvement" of alpine climbing has the potential to foster an environmental ethic.[40] As he puts it,

> [A] bivouac may never have made an environmentalist of anyone, but being immersed in an environment that provokes awe, and that makes obvious your vulnerability to the same elements as wild things are, can profoundly affect your rapport with wildness.[41]

Later, he makes the point more succinctly: "Awe inspires ethics."[42]

While on the surface Kelsey might seem to be privileging sublime landscapes, *A Place in Which to Search* is actually doing something far more complex. As the book's title indicates, Kelsey is a seeker, and his pursuit of all the Winds have to offer challenges the too-easily levied critique that climbing narratives are fundamentally narratives of dominance and conquest. Kelsey engages with important issues surrounding public lands, including use value, the meaning of wilderness, and the impacts of climate change, and he points out tensions between the "mosaic of administrative entities" that comprise the Wind Rivers and the on-the-ground experience of being in the mountains.[43] Much as he delights in reaching summits, Kelsey rarely thinks about his endeavors uncritically, posing questions like "is climbing appropriate in wilderness?"[44] As a part of his contemplations, Kelsey asks whether "another objective of Wilderness management" might be "to produce further great works of literature?"[45] Whether one reads this question seriously or in jest, it invites a consideration of its inverse: how might literature help us think about wilderness management?[46]

A commitment to stewardship runs throughout Kelsey's writing. Long before he published *A Place in Which to Search*, Kelsey wrote what became the standard guidebook to climbing and hiking in the Wind River Range, making several first ascents in the process. Within the climbing community at large, the first person to reach the top of a mountain or specific climbing route earns additional respect—and in the latter case, the right to name the route.[47] This tradition of claiming and naming aligns with Peter Bayers's assertion that climbing culture in the US West "affirms an imperial national narrative that places a triumphant, sublime white manhood at the center of national identity as it is defined against an unpeopled American wilderness."[48] However, in a chapter titled "Antique Pitons," Kelsey describes himself as "a historian and archaeologist" more so than a trailblazer, regularly climbing what has already been climbed in order to document routes for future mountain enthusiasts.[49] As he points out, "for every climber venturing where man has not set specialized climbing shoe, expanding horizons, raising standards, dozens are retracing history."[50] Guidebooks enable such retracing, thereby de-emphasizing the pioneer mentality without significantly diminishing the experience of climbing. For example, Kelsey notes that in following a mentor's route up a peak called the Sphinx, he and his climbing companions "encountered history, while also encountering the here-and-now" of an unexpected climbing challenge.[51] As Kelsey and his friends got older, they developed an appreciation for shorter backcountry excursions and day hikes; he declares, "if we were pioneers, it was as nonperformers of feats."[52]

The terrain in the Winds lends itself to excursions of varying difficulty levels, which means it is not merely a playground for fit young men with a specialized skill set. Although Kelsey was a part of this core climbing demographic in the 1970s, *A Place in Which to Search* is accessible to a much wider readership of outdoor enthusiasts. For example, in Chapter 11, "Through What Does the River Run?," Kelsey is more interested in raising questions about "wilderness" versus "wildness" than in recounting climbing feats.[53] Readers need not be climbers themselves to appreciate his ponderings on these terms.

One of the most striking moments in Chapter 11 is Kelsey's inclusion of a draft version of the *1964 Wilderness Act*, encapsulating his sense of the "spirit" in which one should "wander the Wind Rivers." It reads:

> Wilderness, unlike the earth's landscape dominated by man and his works, is untrammeled by man, a portion of the earth where man himself is a part of the natural community of life, present in humility and with a respect for the natural forces and the humanly unmodified conditions.[54]

The language up to "untrammeled by man" reads much like the definition of "wilderness" in the version of the federal law that was ultimately passed. However, the final version of the act does not recognize man as "a part of the natural community of life," as the draft Kelsey references does, but rather as "a visitor who does not remain."[55] Earlier in the book, Kelsey engages with this latter clause by describing the Winds as a place in which he in fact "*could* remain," but his sense of feeling at home within the wilderness does not translate to an assertion of ownership over it (emphasis added).[56] Instead, Kelsey's sensibilities align with the words "humility" and "respect," which are nowhere to be found in the final version. Kelsey, who does not climb in order to "demonstrate power over nature" but rather to access "emotional identification" through "raw, visceral contact with wild nature," believes that such contact inspires the values present in the draft language above.[57] In their own ways, Robinson and Kelsey each advocate for a contact based approach to land management.

Repeating himself for emphasis, Kelsey states: "Climbing, human artifice though it may be, is undeniably raw, visceral contact with wild nature."[58] However, in the spirit of Robinson's "mountaineering just means glad to be here," Kelsey emphasizes that climbing is not the only viable way to get in contact with the Winds. Chapter 4, "Jones' Columbine," centers around the different ways "becoming increasingly immersed in the Wind Rivers has sharpened [his] seeing skills."[59] "Eventually," Kelsey notes, he began to travel into the mountains "without a rope but with a flower identification guide."[60] He credits friends with less climbing experience but greater knowledge of local plant life for fostering this interest and happily accepts that hiking with naturalists sometimes means setting different goals than he would set on his own or with other climbers. For example, despite failing to reach the summit of Whiskey Mountain one summer, Kelsey and his companions "declared the day a success" because they saw the elusive Jones' Columbine they had set out to find.[61] Additionally, much as Kelsey delights in the active pursuit of certain flowers, geologic formations, or fall colors, he especially celebrates what he "saw unexpectedly" during his time in the Winds and notes that such unexpected encounters are what draw him again and again to the mountain range.[62]

Kelsey compiled some of these unexpected sightings into a list of "noteworthy things," which he wrote for a proposed Forest Service visitor's brochure, and included in Chapter 4. He explains his eagerness to contribute to the brochure in order to help balance out

> the usual litany of Biblical injunctions (thou shalt not camp within 200 feet of a lake; thou shalt pack out thy trash) with prose that would inspire visitors to pack out what they packed in, not because an authority figure was watching, but out of affection for the land.

Kelsey's statement that "preservation begins with engagement with the land and affection, not fear of Jehovah's wrath" is foundational to the overarching message of *A Place in Which to Search*.[63] Nevertheless, preserving a rugged environment means maintaining its qualities as a difficult space for human inhabitation. Spending extended periods of time in the Winds entails dealing with unpleasant weather, facing the threat of wildfires, and accepting losses, variously measured. Kelsey attends to these challenges throughout the book, but dedicates all of Chapter 8, "Bad Weather," to confronting them head on.

"Bad Weather" stands out from the rest of the book from its first sentence: "We are trudging back down a valley."[64] Kelsey not only writes in the present tense, which is how he begins the book's final chapter as well, but also uses the collective "We."[65] Who, exactly, "we" refers to does not become any clearer as the paragraph progresses, nor does Kelsey locate the reader in a time or place more specific than a valley in the Winds. He does, however, describe the severity of the weather in vivid detail, along with the embodied experience of "trudging" through it: "Although our hats are pulled low and our parka hoods drawn tight, the frozen rain stings our cheeks. The wind is inevitably a headwind and drives the pellets horizontally. . . . Dripping snot freezes on our upper lips."[66] After anticipating what making camp in an ice storm will entail and recalling the rappel down the route "we" had set out to climb, Kelsey gives the reader a little more context:

> It does not matter what peak it is, what route we had been trying. This scenario has happened often enough, with only slight deviations from

> the standards script, that memories coalesce. What I am describing is a composite of many retreats in storms.[67]

The combination of ambiguity and specificity calls the reader's attention to the reality that bad weather in the mountains is ubiquitous, while also stressing the ways in which such storms create embodied memories for those who have been caught in them.

After setting the scene with this "composite" description of stormy retreat, Kelsey returns to his usual first-person past tense, but "Bad Weather" continues to surprise. There are a few pages of storm stories—September snows, lightning on exposed ledges, endless days of drizzle—but Kelsey does not dwell in the realm of anecdotes.[68] Instead, he uses bad weather as an entry point for considering two far more existential subjects: the inner state of the mountain-goer and climate change.[69] Even more notably, Kelsey does not tackle each of these topics separately, but rather recognizes their entanglement. Toward the end of the chapter, he muses:

> Learning to accept whatever weather comes your way, accepting the fragility of life and your vulnerability, is essential to immersing yourself in the natural world and thus being a mountaineer. In fact it is one of the (few) traits apologists for mountaineering can use to support the assumption that it builds character. However, I felt simply hollow hiking across the Doubletop Mountain treeline country, under a blue sky, at the end of a pleasant 2009 backpack loop.[70]

After a brief aside to describe Doubletop—"gentle topography with tarns, wildflower meadows, and scattered with whitebark pines"—and express his desire "to be reincarnated as a Doubletop whitebark, a dignified and long-living species," Kelsey returns to the day in 2009 when he "felt simply hollow."[71] He explains that the pine bark beetle had wreaked havoc on the whitebark pine in the area, linking the devastation to climate change: beetles thrive in warmer temperatures, while whitebarks become more vulnerable in drought conditions. A fungus called blister rust and "a century of fire suppression" also threaten the whitebark.[72] The fact that the Doubletop trail is located within the Bridger Wilderness does not mean it is "untrammeled by man." Through calling attention to the limitations of land management, Kelsey asks an implied question: what is the meaning of wilderness within the context of anthropogenic climate change? Neither he nor Robinson offers a clear answer, but their writing encourages getting in *contact* with mountain wildernesses as a way of both thinking and feeling with the question.

As Kelsey makes clear, we cannot go to the mountains to "escape the world's inherent sorrows." However, climbing and hiking in places like the Winds can help us remember "how lucky we are to be there," perhaps fostering larger environmental commitments in turn.[73] *A Place in Which to Search* does not give readers a clear objective to search *for*. The practice or *practices* of searching are what matter.

Notes

1 Chouinard, "Earth Is Now Our Only Shareholder."
2 Chouinard, "Earth Is Now Our Only Shareholder."
3 Gelles, "Billionaire No More."
4 Anson, "The President Stole Your Land," 49–52, emphasis in original. Citing Tuck and Yang, Anson argues that Patagonia's ad also "secur[es] 'settler innocence' by protecting structures of power" (52).
5 Climber and Camp 4 documentarian Steve Roper marks the period as 1947–1971 (Roper, *Camp 4*). Modern climbing/mountaineering in the United States dates back to the mid-nineteenth century, but climbers in the post-World War II period dramatically pushed the envelope on the difficulty of what could be climbed and how, tackling taller and more vertical faces with less and less dependence on artificial aid. Joseph E. Taylor describes the 1972 Chouinard catalog, which announced "a new era of mountaineering," as "a cultural watershed," for its firm assertion of a set of ecological and moral values that would heavily influence climbing attitudes in subsequent decades (Chouinard and Frost, "A Word," 3); (Taylor, *Pilgrims of the Vertical*, 212).
6 Susan R. Schrepfer succinctly articulates the relationship between mountains and wilderness: "Mountains were North America's last, and its first, wilderness," she explains, last, because settlers sought the quickest way through high alpine areas on their way to more easily habitable land, and "first, because it was the mountains that the United States initially and officially established as wildernesses" (Schrepfer, *Nature's Altars*, 1–2).
7 Linck, "Your Stoke Won't Save Us."
8 Cronon's, "The Trouble with Wilderness" is the seminal example of critical engagement with the term "wilderness." Cronon asks, "How can we take the positive values we associate with wilderness and bring them closer to home?" (Cronon, "The Trouble with Wilderness," 87).
9 See Marguerite S. Shaffer for a thorough examination of the early NPS's active promotion of the "nature's nation" ideal in order to frame "the national parks as a system of national assets" (Shaffer, *See America First*, 128).
10 "New Survey Reports on Climbing's Consistent Growth." Given the second Trump administration's heightened attack on both public lands and DEI efforts, continued expansion and diversification of the environmental movement are imperative.
11 @catlikeacat, Instagram.
12 Karlo, "Cat Like a Cat."
13 Honnold and Cahall, "Untelling the Story." I cite the *Climbing Gold* version of these narratives in order to emphasize that the dominant (white, male) climbing culture is beginning to grapple with its settler-colonial legacy.
14 "Devils Tower Voluntary June Closure."
15 Robinson, *A Night on the Ground*, 139.
16 Doug Robinson, personal interview, February 3, 2022. "Scrambling" is an activity that falls somewhere in between hiking and climbing.
17 Robinson, "The Whole Natural Art of Protection," 12.
18 Clark and Hessl, "The Effects of Rock Climbing on Cliff-Face Vegetation." Clark and Hessl conclude that "collaborative stewardship directly involving climbers" is the best way to facilitate "successful management of cliff faces" (714).
19 Robinson, *A Night on the Ground*, 9.
20 Robinson, *A Night on the Ground*, 10.
21 Robinson, *A Night on the Ground*, 119.
22 Robinson's sense of present awareness calls to mind Anna Tsing's insistence on learning "how to look around rather than ahead" (Tsing, *The Mushroom at the*

End of the World, 22). Although Robinson and Tsing engage with precariousness on vastly different spatiotemporal scales—the precarity of an individual on a rock wall versus precarity as "the condition of our time"—they each arrive at their own version of what Tsing calls "arts of noticing," which she suggests can be a form of resistance to future-oriented thinking (20).

23 Robinson, *A Night on the Ground*, 45.

24 McCarthy, "Why Climbing Matters," 170.

25 McCarthy, "Why Climbing Matters," 170–171.

26 Robinson, *A Night on the Ground*, 46.

27 Robinson, *A Night on the Ground*, 77. Poem reprinted with permission from the author.

28 I recognize that "mountaineering" is typically used as a technical term that refers to reaching summits, often those which are high, snowy, and treacherous. However, I think it is worthwhile to introduce a more expansive definition of mountaineering, particularly as there is precedent in the climbing world for putting pressure on the term. For example, in its August 1956 and January 1957 issues, *Summit* magazine ran a two-part series asking the question "What is Mountaineering?". The answers ranged from "the climbing of mountains by the climber's unaided strength" to "getting to know its mountains in all its moods and seasons" (Styles and Washburn quoted in Werner, "What Is Mountaineering?," 6).

29 Robinson, *A Night on the Ground*, 78, 44.

30 Robinson, *A Night on the Ground*, 134. Robinson also describes talus running as "boulder hopping" and "scrambling" (133).

31 Robinson, *A Night on the Ground*, 138.

32 Kelsey, *A Place in Which to Search*, 9. The title for this section is taken from page 120 of the same text.

33 Kelsey, *A Place in Which to Search*, 10.

34 Kelsey, *A Place in Which to Search*, 52.

35 Kelsey, *A Place in Which to Search*, 53.

36 Kelsey, *A Place in Which to Search*, 75.

37 Kelsey, *A Place in Which to Search*, 1.

38 Kelsey, *A Place in Which to Search*, 6.

39 Kelsey, *A Place in Which to Search*, 226.

40 Kelsey, *A Place in Which to Search*, 26.

41 Kelsey, *A Place in Which to Search*, 25.

42 Kelsey, *A Place in Which to Search*, 227.

43 Kelsey, *A Place in Which to Search*, 4.

44 Kelsey, *A Place in Which to Search*, 222.

45 Kelsey, *A Place in Which to Search*, 242.

46 In Chapter 17 of this collection, Barnes and Hamilton also turn to literary texts to think about land management, incorporating poems by Carl Sandburg and Walt Whitman into their engagement with grass and meaning-making at Gettysburg.

47 Writing about successful recent efforts to rename sexist and racist route names, Jennifer Wigglesworth notes "that climbers' activist and advocacy work has not yet dismantled . . . FA (first ascent) naming rights. This tradition continues to wield power" (Wigglesworth, "(Re)naming Routes," 143).

48 Bayers, "Yosemite Climbing Films and the Regeneration of White Masculinity in the American West," 80.

49 Kelsey, *A Place in Which to Search*, 98.

50 Kelsey, *A Place in Which to Search*, 101.

51 Kelsey, *A Place in Which to Search*, 100.

52 Kelsey, *A Place in Which to Search*, 133. Kelsey has a knack for balancing a reverence for both climbing and the Winds with playfulness. For example, he explains that he determined whether an "unroped climb" was Class 2 or Class 3 for his

guidebook based on whether or not his golden retrievers could reach the top (Kelsey, *A Place in Which to Search*, 45).
53 Kelsey, *A Place in Which to Search*, 227.
54 Kelsey, *A Place in Which to Search*, 248.
55 "S.4–88th Congress [1963–1964]: Wilderness Act."
56 Kelsey, *A Place in Which to Search*, 4.
57 Kelsey situates himself in "the tradition of Thoreau and Muir" here (Kelsey, *A Place in Which to Search*, 227).
58 Kelsey, *A Place in Which to Search*.
59 Kelsey, *A Place in Which to Search*, 81.
60 Kelsey, *A Place in Which to Search*, 78.
61 Kelsey, *A Place in Which to Search*, 82.
62 Kelsey, *A Place in Which to Search*, 94.
63 Kelsey, *A Place in Which to Search*, 78.
64 Kelsey, *A Place in Which to Search*, 163.
65 The last chapter (Chapter 12) plays with tense and temporality in differently interesting ways, recalling specific events in the present tense, as though stretching out a continuous present. Rather than disorienting the reader, as Chapter 8 does, Chapter 12 seeks to orient the reader by locating the present moment within eternal time.
66 Kelsey, *A Place in Which to Search*.
67 Kelsey, *A Place in Which to Search*, 164.
68 Kelsey, *A Place in Which to Search*, 169, emphasis in original.
69 Citing *Moby Dick's* Ishmael, who goes to sea "whenever it is a damp, drizzly November in my soul," Kelsey suggests that something similar occurs with climbers, who find themselves seeking "spacious, unpopulated mountains" when their "souls were drizzly" (Kelsey, *A Place in Which to Search*, 171).
70 Kelsey, *A Place in Which to Search*, 173.
71 Kelsey, *A Place in Which to Search*.
72 Kelsey, *A Place in Which to Search*, 174.
73 Kelsey, *A Place in Which to Search*, 175.

Bibliography

Anson, April. "The President Stole Your Land: Public Lands and the Settler Commons." *Western American Literature* 54, no. 1 (2019): 49–62. https://dx.doi.org/10.1353/wal.2019.0019.

Bayers, Peter. "Yosemite Climbing Films and the Regeneration of White Masculinity in the American West." In *The Routledge Companion to Gender and the American West*, edited by Susan Bernardin. Routledge, 2022.

Chouinard, Yvon. "Earth Is Now Our Only Shareholder." *Patagonia*, September 14, 2022. https://www.patagonia.com/ownership/.

Chouinard, Yvon, and Tom Frost. "A Word." In *Chouinard Equipment Catalog*. Sandollar Press, 1972.

Clark, Peter, and Amy Hessl. "The Effects of Rock Climbing on Cliff-Face Vegetation." *Applied Vegetation Science* 18, no. 4 (2015): 705–715. https://doi.org/10.1111/avsc.12172.

Congress.gov. "S.4–88th Congress (1963–1964): Wilderness Act." September 3, 1964. https://www.congress.gov/bill/88th-congress/senate-bill/4/text.

Cronon, William. "The Trouble with Wilderness; or, Getting Back to the Wrong Nature." In *Uncommon Ground: Rethinking the Human Place in Nature*, edited by William Cronon. W.W. Norton & Company, 1996.

"Devils Tower Voluntary June Closure: An Opportunity to Show Respect." *Access Fund*, May 1, 2024. https://www.accessfund.org/latest-news/news-and-events/news/climbers-honor-the-june-closure-at-devils-tower.
Gelles, David. "Billionaire No More: Patagonia Founder Gives Away the Company." *New York Times*, September 14, 2022. https://www.nytimes.com/2022/09/14/climate/patagonia-climate-philanthropy-chouinard.html.
Honnold, Alex, and Fitz Cahall, hosts. *Climbing Gold*. Season 1, episode 9, "Untelling the Story." June 18, 2021. Podcast, 48 min., 39 sec. https://www.climbinggold.com/episodes/chapter-09.
Karlo, Kathy, host. *For the Love of Climbing*. episode 29, "Cat Like a Cat." August 1, 2021. Podcast, 1:00:48. https://open.spotify.com/episode/0C3exvKqsJHR2eqfDLV2Gk?si=cad57cb906a242ca.
Kelsey, Joe. *A Place in Which to Search: Summers in the Wind Rivers*. Black Canyon Books, 2016.
Linck, Ethan. "Your Stoke Won't Save Us." *High Country News*, May 14, 2018. https://www.hcn.org/issues/50.8/recreation-your-stoke-wont-save-us.
McCarthy, Jeffrey Mathes. "Why Climbing Matters." *Interdisciplinary Studies in Literature and Environment* 51, no. 2 (2018): 157–174. https://doi.org/10.1093/isle/15.2.157.
"New Survey Reports on Climbing's Consistent Growth." *Climbing Business Journal*, September 23, 2022. https://climbingbusinessjournal.com/new-survey-reports-on-climbings-consistent-growth/.
Robinson, Doug. "The Whole Natural Art of Protection." In *Chouinard Equipment Catalog*, Sandollar Press, 1972.
Robinson, Doug. *A Night on the Ground, a Day in the Open*. Mountain N' Air Books, 1996.
Robinson, Doug. Personal Interview. February 3, 2022.
Roper, Steve. *Camp 4: Recollections of a Yosemite Rockclimber*. The Mountaineers, 1994.
Schrepfer, Susan R. *Nature's Altars: Mountains, Gender, and American Environmentalism*. University Press of Kansas, 2005.
Shaffer, Marguerite S. *See America First: Tourism and National Identity, 1880–1940*. Smithsonian Books, 2001.
Taylor, Joseph E. *Pilgrims of the Vertical: Yosemite Rock Climbers and Nature at Risk*. Harvard University Press, 2010.
Tsing, Anna. *The Mushroom at the End of the World: On the Possibility of Life in Capitalist Ruins*. Princeton University Press, 2015.
Werner, Louise Top. "What Is Mountaineering? Conclusion." *Summit Magazine* 3 (January 1957): 6–7.
Wigglesworth, Jennifer. "(Re)naming Routes: A Tale of Transformation in the Outdoor Rock Climbing Community." In *Gender, Politics, and Change in Mountaineering*, edited by Jenny Hall, Emma Boocock, and Zoë Avner, 143. Springer Nature, 2023.

20 Hueco Tanks

Envisioning Indigenous Space and Public Lands

Robert Gunn, Andrea Everett, and Alex Mares

Mapping Routes: Pictography, Bouldering, and Indigenous Sovereignty

Near the junction between East Mountain and West Mountain at Hueco Tanks sits a round boulder balanced upon a plinth of stone, a natural landmark shaped by wind and water and time. Approached from the south, it is a visual anomaly that juts out against the rising backdrop of East Mountain at Hueco Tanks, itself a geological anomaly within the Chihuahuan Desert, about 25 miles from downtown El Paso, Texas. At this place—perhaps 800 years ago, perhaps 1,500 years ago—a traveler brought with them knowledge of a sacred tradition evoking Tlaloc, the Rain God, from central Mexico and, lying on their back, fused that knowledge with dark pigment onto the underside of an outcropping of syenite porphyry. The hidden location of the Tlaloc pictograph accounts, in part, for the extraordinary vividness of its imagery, as it has survived hundreds of years shielded from the weather. Its hiddenness also speaks to the occult nature of the vast visual array of sacred things that reside here. Unless you have been shown or told that it is here, you would be unlikely to discover it. But once you know how to look for it, the purport of this rock formation as a sacred landmark is unmistakable, accessible to eyes that have been trained to see and recognize in its proximity to the hidden figure of the Tlaloc the joining of human intentionality and the natural environment at the axis of a higher reality.[1]

Like my fellow coauthors, I am certified as a volunteer pictograph guide at Hueco Tanks, where we lead tours of fragile environments and sensitive cultural areas, including "guide only" restricted zones like Tlaloc pass. Unlike my Indigenous coauthors, who speak so eloquently elsewhere in this chapter of their respective relationships to Hueco Tanks, mine is a settler relationship to Hueco Tanks. I draw lessons from them both in my attempts to think conscientiously about what that difference of relationship entails, and how I should approach and understand my role as a guide who is also always a visitor. I also do my best to draw lessons from Hueco Tanks itself, which offers lessons in humility in the presence of sacred matters and the human mysteries of deep time. Here is one thing I have learned: pictographs offer lessons in the

DOI: 10.4324/9781003581185-25

sovereignty of sacred spaces.[2] For Indigenous persons who are initiated into the nature of their traditional purposes and realities—by virtue of knowledge inheritance, community traditions and ceremony, rituals of storykeeping, or other forms of kinship and cultural affiliation—the pictographs at Hueco Tanks may announce proximity to sacred presence and ways of knowing that are often inaccessible to, or are simply not for, outsiders. To put this in a slightly different way, pictographs reveal only as much as one is prepared to perceive, and no more. What is inviolable about the sovereignty of pictography rests at the threshold between what may be observed, experienced, and re-signified as living cultural practice and that which remains unknown or is perhaps unknowable.[3]

Ultimately, the natural sovereignty of Hueco Tanks is indisseverable from the cultural sovereignty of the Indigenous people who have held it as sacred for millennia. The recent advent of sport climbing and bouldering at Hueco Tanks, however, demonstrates that sovereign and sacred places may nevertheless be deeply vulnerable to willful acts of degradation, defilement, and disrespect. Sport rock climbing at Hueco Tanks was rare in the 1950s and 1960s, but accelerated in the 1970s and 1980s with the growth of its reputation as a premier destination. Hueco Tanks is identified very closely with the origins of bouldering in particular and, since the 1980s, has been heralded by the climbing community as offering the best bouldering opportunities in North America, if not the world. Climbers have mapped their presence in every corner of Hueco Tanks. Alongside the same trail leading to the Tlaloc figure above, less than 100 yards north of Tlaloc Pass, another stone between East and West Mountains rises from the earth. Rock climbers today giddily call it, "Raging Pecker." John "Vermin" Sherman, author of the most comprehensive and widely -used climbing and bouldering guide of Hueco Tanks available, describes this rock as "a 50 foot long phallic-shaped boulder pointing towards the Spunk Rocks 60 yards to the northeast."[4] Likening a natural rock formation to an erect penis, Sherman invites climbers and readers alike to imagine that rock formation in an act of perpetual ejaculation onto the mountainside 60 yards away, an obscene fantasy that is reinforced by the names of the climbing problems and routes he maps onto the trajectory between. "Three Pulls and It's All Over," "All That Jizz," "Spunk," and "Spunky's Playhouse," plus six more climbing problems whose names are cartoonish synonyms for a penis.[5] The name given to another climbing route in this area, "Kinjite"—for the 1989 Charles Bronson movie, *Kinjite: Forbidden Subjects*, the plot of which revolves around male rape fantasies and the sexual exploitation of young girls—offers a coy acknowledgement of what is more deeply at work here in this naming pattern in the form of a joke that serves also as misogynistic provocation. These subjects may be "forbidden," but here, in a quasi-pornographic "playhouse" in which male fantasy is made material, nothing is sacred from trespass. The macho posturing of this stake-claiming is overt and hostile, as is the gendering of the landscape as a type of female embodiment available for sexualized domination and defilement. This is in fact a major theme of Sherman's guide. There are perhaps 40 climbing routes or bouldering problems which reference male anatomy or sexual acts, as above; but there are *more than 200* names for landforms or bouldering problems that sexually objectify girls or women, explicitly figure female genitalia or breasts, or identify physical climbing actions on rock formations with sexualized acts involving female anatomy.[6] The collection, publication, and distribution of these names through multiple editions of Sherman's *Hueco Tanks: Climbing and Bouldering Guide* represent a key episode of ideological consolidation for the institutionalization of misogyny in the sport through one of its most authoritative and foundational documents.[7]

But if John "Vermin" Sherman, the man and his books, is now a kind of storied institution unto himself in the rock climbing world, the outlaw image he has coveted so carefully over the last 40 years is deeply threaded into the edifice of that institution and the modern history of rock climbing itself.

The short history of climbing and bouldering at Hueco Tanks, specifically, is a model of the processes of settler colonialism whereby acts of illegal occupation and lawlessness become laundered through time and mythification as foundational acts of self-legitimation. In a retrospective recounting the history of climbing at Hueco Tanks, Todd Skinner offered this recollection of the 1980s' practice of illegally bolting the rocks to establish new climbing routes at night:

> Half the reason we did anything was that it was forbidden. We had scouts. We had relays. It was like a war where nobody got killed. One night, some unknown person was drilling *Tarts of Horsham*. The light went on, which meant "quit drilling." Then the light went on and stayed on, which basically meant "every man for himself and nobody get caught." Then the unknown person dropped his jumar, and had to rappel into a group of rangers and fight his way out. The end result was harmless. They make rules; we break 'em.[8]

What stands out here is not only the blithely antisocial rejection of any sense of duty to operate by standard park prohibitions against destructive activity but rather the way figures like Skinner and Sherman viewed their illegality as a form of cultural role play ("it was like a war where nobody got killed"), complete with clandestine operations, mock battles, and scripted lines ("every man for himself'). But these actions are of course not war; nor is self-serving vandalism in the name of sport any kind of principled activism or civil disobedience. They belong to the same space of individualist male fantasy as the juvenile and degrading names of the more than 1,800 climbing problems at Hueco Tanks, a kind of gleeful, theatricalized reenactment of nineteenth-century squatter colonialism in which they may live out some of the central tropes of the Hollywood outlaw hero.

The upsurge of activity at Hueco Tanks in the 1980s brought litter, destruction of fragile ecosystems, and permanent damage and alteration of the sacred rocks of Hueco Tanks in the forms of drilled bolts, chipped and broken rock faces, and lasting adherence of chalk to stone. In response, Texas Parks and Wildlife closed the park temporarily in 1992 for damage assessment, closed permanently some climbing sites, and then adopted new Public Use Plans in 1998 and 2000 that significantly curtailed climbing access in the backcountry while implementing ambitious educational and guide-training programs. The climbing community, who once wore proudly their antiestablishment scorn for rule and law, pivoted quickly to rebrand themselves as aggrieved and legitimate stakeholders whose rights had been violated. A bellwether signaling this shift at this juncture was the emergence of the Access Fund, an outside lobbyist and legal action organization which entered the public debate in 2000 proclaiming "clear institutional bias against climbers at Hueco Tanks."[9] The success of this lobbying strategy was evident in the drafting and publication of a new mission statement for Hueco Tanks State Park in 2003, which for the first time granted explicit recognition to climbers as a legitimate "user group" and endorses a new definition of the nature of Hueco Tanks itself: "The site is significant or recognition as one of the premier bouldering locales in the world."[10]

These developments illustrate concisely how the processes of settler colonialism short-circuit, and ultimately corrupt, the politics of recognition—the latter, long heralded in progressive politics and theories of political liberalism as a prerequisite to social justice and conditions of equality—to become a tool of legitimacy laundering, converting past acts of criminal appropriation into the foundation of a stakeholder identity that claims its own rights must be respected.[11] Far from transforming the political dialogue or ethical basis of that community, recognition of historically bad actors instead undermines the legitimacy of the model of recognition itself. Consider the following statement from an unnamed "veteran Hueco Tanks climber" in 2003, the same year that recognition was conferred upon climbers in the Hueco Tanks mission statement, which reflects a newly domineering sense of title:

> Only climbers can really sense when the essential climbing energy of a place is still intact, or has been sucked away by the disrespectful presence of land managers and culture junkies. Imagine the insult the rocks have to endure—the people would walk to them, and instead of climbing, ignore the poor boulder to look at graffiti that was painted on by uncaring aborigines thousands of years ago.[12]

This style of arrogant hyperbole turns on the same wry whiff of irony that animates the hidden joke behind the bouldering problem named "Kinjite," above—strategically enlisting the plausible deniability of a joke, but in a manner that only underscores the sincerity and depth of the insult intended, in a claim that denies the recognition of the legitimacy of Indigenous space while reasserting the supremacy of settler territorialization of Hueco Tanks in the form of climbing's misogynistic routes and topographies.

An analog to this kind of misogynistic epistemology of land and domination may be found in a famous scene in Cormac McCarthy's *Blood Meridian*, which takes place at Hueco Tanks. In chapter XIII, the fearsome, seven-foot albino villain known as the Judge surveys hundreds of pictographs, specimen book in hand.

> They were of men and animals and of the chase and there were curious birds and arcane maps and there were constructions of such singular vision as to justify every fear of men and of the things that are in him.

Moving with assurance among these images, he carefully copies select examples into his book.

> When he had done and while there yet was light he returned to a certain stone ledge and sat a while studied again the work there. Then he rose and with a piece of broken chert he scappled away one of the designs, leaving no trace of it only a raw place on the stone where it had been. Then he put up his book and returned to the camp.[13]

In a novel replete with disturbing depictions of violence against human beings, this cryptic act of silent and methodical defacement still stands out with a cold and eerie malevolence. In the Judge, McCarthy's malevolent counterpart to Melville's Moby Dick as an avatar of monstrous whiteness and supernal aggression, the novel mirrors for readers the end-stage distillation of the logic of Western expansionism as a technology of violence simultaneously against people and against the earth itself. More specifically, in highlighting the presence of the Judge's specimen journal, McCarthy depicts the nineteenth-century practice of ethnological specimen collection as a perversely zero-sum epistemology, a project that constructs knowledge mastery through a process of Indigenous appropriation and simultaneous cultural erasure. Readers of McCarthy typically place this moment in conversation with others in which the Judge's journal features prominently, particularly the episode which proclaims to Toadvine the psychopathic logic behind his specimen copying and collecting: "that which exists in creation without my knowledge without my consent."[14] What interests me here, instead, is what happens next. In McCarthy's ghastly picaresque, like all picaresques, episodic sequence is a compositional tool of meaning-making. Immediately following the bounty hunters' night at Hueco Tanks is a massacre. Three days roaming brings them to a band of Tigua along a river, and, senselessly and mercilessly, they attack. In McCarthy's telling, it is young women tending fields and children who are the first victims, but none survive. It is older Tigua women who discover the carnage. To reinforce the connection between this event and the Judge's effacement of the pictograph at Hueco Tanks, McCarthy recurs to the same theme of erasure to describe both:

> in the circuit of a few suns all trace of the destruction of these people would be erased. The desert wind would salt their ruins and there would be nothing, nor ghost nor scribe, to tell to any pilgrim in his passing how it was that people had lived in this place and in this place died.[15]

In drawing these two moments together, McCarthy implicitly equates them and asks us to recognize that the violence against Hueco Tanks correlates to violence against Indigenous persons, Tigua women specifically.

—**Robert Gunn**

A Pueblo Woman's Rage

Spirits and Connections of this Cultural Landscape

Indigenous maps differ from settler-colonial maps, which aim to own and conquer. Colonial maps impose power structures on Mother Earth, converting her into an inanimate commodity to be exploited, a resource to be extracted. Indigenous maps seek to illustrate relationships; they are communal and rooted in relationality. The caves, rocks, and boulders at Hueco Tanks are not commodities without a soul; they are sacred because they are physical places that resonate through time and space, holding all our spirits woven like a spider's web, repeating patterns and connecting us to our ancestors and their primordial knowledge of migrations, places, and maps. A task made more difficult by people seeking to "recreate" (recreational activities) in sacred spaces.

My Pueblo culture is land-based. I have a deep connection to the energy and spirit of what is now known as Hueco Tanks. As a child, I was taught to see myself as being a part of the land, and I still do. When I connected with the living spiritual beings emanating from the rocks and boulders, as well as the desert plants and animals, I began to understand the importance of reverence and offerings. I learned how esteemed this place was to my ancestors and made a vow to protect it. I have always maintained my Pueblo worldview and interpreted Western science through an Indigenous lens. Like other geographical markers on the land, Hueco Tanks is a source of power on this Earth. This same power lies within us, allowing us to develop spatial reasoning and map landscapes in our minds, practice, and reconnect to the innate mental maps that we are born with.

Knowing this allows me to view the land from a directional perspective, honoring the concepts of cardinal directions and celestial spheres. These celestial spheres represent our consciousness and our inner-deepest knowing. Hueco Tanks and the deities on the rock surfaces serve not only as points of directionality aligned with the Earth's cycles, celestial patterns, and phases but also as ancestral teachings that we must interpret as a way forward to reimagine and realign ourselves in spaces that never asked to be controlled and limited through colonial interpretation.

A Pueblo Woman's Rage

I am Tigua, from Ysleta del Sur Pueblo, a federally recognized Native American sovereign nation. Our original homelands are located at Quarai Pueblo in east-central New Mexico. During the 1680 Pueblo Revolt, about 400 Tigua were captured by the Spanish and forced to walk nearly 400 miles south. In 1682, the Tigua settled and established the Pueblo of Ysleta del Sur (YDSP). Over time, ongoing colonization, land encroachment, and cultural erasure continued to diminish our tribal land. Our access to natural systems and important cultural sites, such as Hueco Tanks and the Rio Grande, which once sustained us and where we hold and participate in ceremonies, is now limited.[16]

I often hear in Indian Country, "[w]e live in two worlds." Still, as I have grown older, I realize I am not separate from the world my ancestors experienced; we share the same world, where time and space bend to keep our souls and spirits forever linked. The distinction is that we live within systems meant to control and subjugate us. This control begins with the controlling of the land and her gifts and slithers into our bodies, spirits, and minds.

My body carries a deep rage. I am only 4 feet 11 inches tall, yet I carry the anger of seven generations of Pueblo women. Sometimes, it becomes too much for this small container, and it spills over when injustice overwhelms me. I have come to understand that my rage is sacred; it energizes me, sustains me, and keeps me moving, but at times it has infected me and turned into poison. Most of my life, I have never felt like I truly owned my own body. I wonder if the land, especially sacred places like Hueco Tanks and our river, "Pehla Usla" (Big River/Rio Grande), feels as oppressed as I have throughout my life. When I sit with the land, I do not feel rage; I feel her sorrow. I sense her labored breathing, her cries, and her sense of being used. I compare it to a victim of sexual assault, and as one myself, I trauma bond with her, and our mutual dependence connects us. The anger that is always present in my body, as we navigate late-stage capitalism, as these systems of harm cling tighter to avoid their collapse, needs a release or a transformation. Nevertheless, we remain in a liminal space filled with uncertainty, not knowing when healing will arrive.

Nevertheless, my anger holds transformative potential. Like fire, it is my medicine, offering a path toward renewal. What does it mean to intentionally ignite the fire within us instead of fighting the fire inside? How can we ignite small, controlled fires within ourselves and in our communities, so that the nutrients left behind can nourish us and restore our spirits? Burning away the dead parts of ourselves, charred remains giving way to vibrant life and enriching us and our lives.

History of Land Grabs and Extraction of Spirit

For Indigenous people, those who are still practicing cultural ceremonies, speaking their language, and living a life rooted in indigeneity, the land is not a liquid source of wealth; the land is the most fundamental relationship (the Western mindset would say the land is the most basic resource). Hueco Tanks is evidence of that. Through the creation of colonized maps, land deeds, and false treaties, the autonomy and self-reliance of the Indigenous people of Turtle Island were disrupted. Cultural genocide was furthered by not allowing access to cultural power points in the landscape. However, they did not eradicate those practices or beliefs; they merely lay dormant for a short time. Hueco Tanks is a repository of wisdom, containing intricate instructions coded in ancestral knowledge conveyed through geography, celestial alignment, ceremony, and living spirit beings. Despite these hardships, the Tigua have continued to gather natural elements—such as water, plants, and minerals—to sustain our spiritual practices and maintain our connection to the river and sacred cultural landscapes, even as environmental changes have made this increasingly difficult.

Time and time again, white politicians, businessmen, and today, this includes others who have adjacency to whiteness sought to use and exploit us, entice us to steal our land, what they see as "resources," and strip us of our relationships to the land and with each other. In the case of Hueco Tanks, this also holds historically, as evidenced by its takeover as a state park. Attorney Tom Diamond inadvertently reveals that bigoted politicians created reasons to push their personal projects and interests through back-room deals that purportedly served the "best interest" of the Tigua. However, the Tiguas' relationship with Hueco Tanks was used as leverage to build a county road, not to provide a means to protect Hueco Tanks and the spiritual relationship that the Tigua and other Indigenous people had/have with this sacred place. As asserted in the 2015 book, *Moon Spell*:

> After recognition, Barney Weiland approached me about getting a county road paved to the Hueco Tanks area. I talked to County Judge Coldwell and he was all for doing it provided we had a good reason. The best reason we could think of was a State Park at Hueco Tanks. The initial plan was to have Weiland convey the land to the Tribe and have the Indian Commission make the request to the county for acquisition of the land as a park. We decided we needed support from the Texas legislature to make the plan work and hit a road block with Joe Christy who was for the idea but only if we conveyed title to the State itself and not to the Tribe. We finally succeeded after reaching an agreement with the Parks and Wildlife Department that the Indians would run the Park under contract with the State. The State of Texas, after a couple years reneged on their commitment to the Indians run the park and took over management themselves.[17]

Reverence versus Recreation

We are taught to enter these spaces with humility, keep our voices low, and respect the spirit beings that are always present. Conversely, we hear the loud noises of visitors cheering each other on in areas they have named vulgarly and disrespectfully, as they seek to climb and conquer. The rowdy sounds of bodies collapsing onto crash mats reverberate through the ground and echo throughout this hollow space. My elders have taught me that this place acts as a filter, where we go to the land and ask for relief from our pain, sorrow, and struggles. We are told to leave those burdens there so we can return home renewed and regenerated, allowing us to uphold our traditions and live in a good way. Hueco is a place that has cradled and healed me time and time again. However, it is hard to sit in prayer when your church is treated as a jungle gym, and the noise drowns out any sense of peace and release.

At places like Hueco Tanks, it is crucial to prioritize reverence over recreation. It is also essential to challenge the colonizer's mindset, which fuels their passion for the sport—specifically climbing in the case of Hueco Tanks—and clouds their understanding of the spaces' historical, political, and cultural significance, as well as the cultural significance of the spaces they temporarily occupy.[18] These mindsets, whether conscious or not, sustain violence against the land and, consequently, violence toward Indigenous people, especially Indigenous women. Pueblo women have had their roles and connections to the land violently stripped from them—whether through stripping us of our matrilineal roles of overseeing the land and our homes or denying us our roles in birth, puberty, and death ceremonies. This extraction of the matriarch spirit from the land and ceremony has severely harmed Indigenous bloodlines, causing intergenerational trauma in our bodies, disconnecting us from our intuition and sacred gifts, which has led to a spiritual disconnection for our people, as all souls pass through the mother. I can only speak to my personal experience as a Pueblo woman—the pain my body carries; the messages my ancestors send me; and my reconnection to the land that helps heal me, my family lineage, and our Mother Earth. I now understand that the only path forward toward healing is through rematriation, the path of the ancestral grandmothers.

The Only Way Forward Is Rematriation

As we move forward and support one another, it is essential to rebuild our communities and rematriate the land, with women and youth leading the way to liberate and restore the land, rivers, and one another. The path forward will be rooted in the relationships we build, repair, and reconnect with. We must create and restore Indigenous and cultural networks and ecosystems that embody the strengths of our ancestors, respect our mutual relatives, align with the divine design of the universe, and follow the original instructions of our ancestors.

The values and teachings of our ancestors will guide us through the years to come. We need to stay grounded and committed to uplifting, caring for, teaching, and forgiving one another. Colonialism resulted in competition with one another, valuing individualism, personal wealth, and success over communal values. Nonetheless, nothing could be further from the truth.

To me, rematriation is a process of becoming and embodying the true path of our Indigenous souls—listening to the fundamental lessons of our ancestors, decoding, and returning to the mental maps we were born with. Rematriation embraces restoring balance within our communities and our bodies, which in turn heals the land and fosters a relationship with Mother Earth. This time and place call for the return of matriarchy. Suppose we represent and commit to moving forward with rematriation as a structure that will replace the crumbling Western systems of harm. In that case, we can call upon the grandmothers and begin to restore balance in our communities and to the land.

Recently, other Tigua women and I have formed a Tiwa Women's Circle (T'ai'u Hlioran Hurini). The Tiwa Woman's Circle seeks to complement outcomes that preserve and protect cultural land heritage by taking essential actions to foster community and ancestral consciousness, address historical trauma and healing, build matriarchal strengths, and rebuild cultural lifeways for the well-being and food security of all Tigua. T'ai'u Hlioran Hurini is actively building soil health, food security, and food sovereignty projects needed to build resilience through a return to traditional farming practices. The T'ai'u Hlioran Hurini (Tiwa Woman's Circle) garden is the quintessence of rematriation and the restoration of spiritual connections to the land, honoring ancestral knowledge and practices.

This project, which started in late 2024, has acquired 0.77 acres of land and two medicine (greenhouses) houses. We have completed one successful planting season, during which we grew various types of corn, squash, gourds, amaranth, tobacco, chiles, melons, and tomatoes. The most valuable growth has been in community and relationships. We honor sacred teachings by living the original instructions of our ancestors through prayer and offerings. It has not been without obstacles. Conflict resolution and harm reduction have been foundational in advancing the priorities of rematriation.

Earlier, I spoke about the cries of sorrow that I hear from the land when I sit and contemplate with her. Although true, it is not the only thing that whispers to me. There are places within Hueco Tanks where the songs of women reverberate. These ancestral women once prepared food and medicine together in the ancient grinding holes. You can sit in the same areas—like a rock ledge overlooking the beautiful landscape or a cave providing shelter from the desert heat—where a spirit friend painted directions with mathematical precision and intentional placement, telling us exactly where to make our offerings so that we live as the Creator intended and acquire the ancestors' good blessings. Rematriation involves returning to the cyclical nature of our lives, rather than adhering to hierarchy and power dynamics. The circle of life represents our kinship with the land and relations to one another through the seasons, climate patterns, observations, and practices.

If we are to truly honor the process of rematriating Hueco Tanks and our world, it begins with how we think, envision, speak about, and the language we use when handling this land and sacred space. Culture and language are intertwined and cannot be separated. Patriarchal culture and language emphasize the authority of fathers and men. However, unlike matrilineal cultures, this authority is often upheld violently through control, dominance, misogyny, and fear. At times, male dominance subjugates and maintains supremacy and exclusivity through the creation of generic user times. Terms that have been used in the past, such as "climbing season" (or "busy season"), promote and favor specific user groups. Whether intentional fragment or not. During these "busy seasons," the climbing groups flood the sacred grounds, and a revered place becomes a boisterous playground.

Both activities are sports, not a culture, and have existed for less than 45 years (technical climbing) and under 35 years for bouldering at Hueco Tanks. The reality for us, as Indigenous people, is that technical climbing and bouldering were introduced, promoted, and supported not by locals or Natives but by a predominantly white institution. Displacement, access, distance, and user preference make it difficult to be on the land and practice our culture. People and cultures that are bound with sacred places and lands are connected through living continuity with our Mother Earth. Because the Tigua were removed and separated from their traditional lands, relations with Hueco were forcefully interrupted.

The call to decolonize and rematriate spaces like Hueco Tanks and to adopt matriarchy as a system to replace extractive practices is growing louder. Ceremony opens a door to the knowledge, maps, and sacred roles of the matriarchs. We have begun our return and ascent to our rightful roles on the land as Pueblo women, for we are the safe vessel to hold the knowledge and messages that the ancestors and living spirit beings are trying to convey and pass on.

—Andrea Everett

Alex Mares—Hueco Tanks Interview with Robert Gunn and Andrea Everett September 4, 2025

This interview has been redacted and edited for length.

[*Robert Gunn*] *Alex, I wonder if you might begin just by introducing yourself and telling us about your relationship to Hueco Tanks.*

[*Alex Mares*] Yá át ééh, my name is Alex Mares. [I'm of Diné descent and Mexican American descent. I am born "to" the Mexican People, and born "for" the Mexican Clan. My maternal grandfather was from The Cloud People and my paternal grandfather was from the Corn People Clan.] I'm the former lead ranger for Hueco Tanks. I worked there for nearly 15 years and lived there with my family for about 10 years. My relationship with Hueco Tanks began prior to it being, quote unquote, owned and controlled by the State of Texas. That's a pretty recent happening or event. So I was born in 1966 and the State of Texas doesn't take possession of that place until after my birth, about four to five years after my birth. So my first connections to Hueco tanks, my first interactions actually happened as an infant. I'm taken out there by my family, by my parents. My umbilical cord is literally buried out there because this is my mom's homeland. This is my mother's homeland area. In turn, my daughter and my granddaughter and soon my grandson's umbilical cords will be buried out there as well. It is. It's more than a symbolic act in why we do that. But it serves to tie us, to root us, to connect us and remind us of where we come from and who our mother is and where our mother is and where our grandmothers are.

There's times where, as a storyteller, I try to relate events that occur out there, natural events, recurring events, cycles, what in Western culture oftentimes are called natural phenomena. And it's really hard sometimes to convey because unless you experience it, unless you're there, unless you see it, go through it, live it, it's really kind of hard for people to conceptualize it and relate to it. But I try and I do it anyway. And I think it's important because especially now in the reality we live in, people are becoming more and more detached from nature and places like Hueco Tanks. It is sacred. It isn't sacred because we say it's sacred. It is sacred intrinsically, inherently. It's sacred before we, as even human beings, even come into the picture. And thus it deserves respect. Navajo Way, every living thing is sacred. . . . The concept of animate and inanimate is foreign to us. That's new to us. Everything is living things that have life and spirit. and Hueco Tanks is definitely one of those places, you know. All land, all of our Mother Earth is sacred, but there are particular spots or points on the earth that have special significance and that have special roles in our origin stories, in our migration stories, important events, powerful events, sometimes even sad events, which all add layers of sacredness to those points on the earth. And it's important that all of us recognize that, not just us as Indians, but all humanity, that we all recognize that.

[*Andrea Everett*] *What should people understand about the cultural and environmental significance of Hueco Tanks for Indigenous people of this region? And how does Hueco Tanks fit into the larger spiritual geography of the Southwest?*

[*Alex Mares*] Hueco Tanks to me is a Rosetta Stone . . . that really hasn't been discovered or acknowledged by non-Natives. But when Natives go there and people that know their individual cultures, religions, spiritual beliefs and practices, when elders go there, and I've had the privilege, I've been blessed to be in the company of older Plains people, older Hopi people, older Pueblo people, older Navajo people, older Apache people. And when they see the images out there, and when they see the images in context and in relation to what's on the land and where they are placed and where they are located, they get it. They understand it. They can relate to it. They can interpret it. Whereas somebody outside of that reality, that community, they see rocks, they see trees, and they see images.

The late John V. Davis said, at the pyramid sites in Mesoamerica, Mexico, and at the big Pueblo ruins [such as Mesa Verde and Chaco Canyon], he says, you might have one, two, maybe 3 periods of archaeology there, but essentially it's the same group of people over time. Not a whole new group coming in, this kind of thing. He says, at Hueco Tanks—and this is a very powerful statement—at Hueco Tanks, every known culture that has traveled through, come through, or lived at in the southwest has left evidence at Hueco Tanks. And, you know, that's a powerful, heavy statement. And on and on and on, up through the Spanish Entrada, Mexican period and Anglo, modern Anglo-American period, they're all there. They've all left their mark and record there.

[*AE*] *How has access to local Indigenous peoples for traditional purposes, spiritual pilgrimage and ceremony changed over time? How have you seen it change over time?*

[*Alex Mares*] In the late 60s, early 70s, and even into the 80s, there was more of a presence. And it's funny because our presence wasn't encouraged. Our presence wasn't welcomed and invited. We felt, as outsiders, we felt unwelcome. And it wasn't just because all the rangers were white and male. You know, they could have been very different. The problem was the mindset and the institutional racism, bigotry even, you could say. And the ones that would try to be fair with us and welcome us and whatever, they were few and far between and they didn't last. And those of us born, mid-60s, early 60s, and the 50s, we remember a time when it was not owned and controlled by the State of Texas, when it was not a State Park. Okay. And as long as we're around and as long as we're alive and we remember that, that's a powerful thing . . .

So on the good side of things [are] the Apache and the ceremonies that we brought back—the Gáhe. . . . The [first] Gáhe ceremony we did back in early 2000s, yeah, late 90s, early 2000s. As near as we can tell, that ceremony had not taken place at Hueco Tanks in over 120 years. So several generations had passed. So to bring that back was powerful and important, and to do it with the cooperation, the blessing, the participation of Texas Parks and Wildlife, that was meaningful. That was important. I just don't know if it could have happened a decade before. . . . And so then we had three subsequent [ceremonies]. The last one was last year, last November. And so we had a total of four. And every time some of the same and many new people came, people of all colors. They were welcome and invited. And again, following that spirit of that place, for human beings to go there to respect, pray, bless, leave blessings and be blessed, make offerings, leave whatever's burdening them, harming them, hurting them, worrying them, stressing them, and do it together and re-create ourselves. You know, that Gáhe ceremony really, really epitomizes and really sends that message home to everybody. And just because you put it on paper doesn't make it so. But at least now we've got documents affirming, reaffirming, supporting our position, our connections, our reality, our relationship to that place.

[*AE*] *What lessons might the repatriation of Uluru in Australia to Aboriginal management have for Hueco Tanks and the Western model of resource management?*

[*Alex Mares*] I would say . . . we don't have to reinvent the wheel. We have some really good templates and examples. . . . And you know, humanity, people are a funny thing. They did a three-year heavy marketing interpretive program to get people on board [with the restoration of Indigenous Australian title to and management of Uluru]. They put out flyers, they put out banners, they put out posters, they put out leaflets, they did radio, they did TV for three years, letting everybody know in advance. Australians and the outside world: in three years, you're no longer going to be able to hike up to the top of Uluru. You're no longer going to be able to climb around these areas or bolt or drill. . . . It's coming to an end. And the thing about people is if you're fair with them, if you communicate with them, if you inform them, if you educate the most people, guess what? They do the right thing. They on their own, they on their own began to not come, even by year one. They on their own began to come and not hike or go up to the top, on their own. By year three, it was a little handful of hardcore holdouts.

I'm not a fan of that whole "management," "resource management" [terminology]. We don't manage a damn thing. The earth manages itself and she's been doing fine for billions of years before we came along. We, if we're going to manage anything, we need to manage ourselves and we need to manage others. We manage ourselves in the same way we manage others. We manage ourselves by educating and forming ourselves. We manage ourselves by being responsible. We model that. And then that in turn educates and informs others and helps manage them and then manage themselves. And if we're all doing this, then the land has nothing to worry about. Then . . . we minimize or we have none of that negative human footprint or impact. Right now, we can't say that. Right now, we cannot say that out there. There's damage and impact right now, in real time. And it's not always what you're doing out there.

Sometimes it's what you're not doing. Every day that there is not a collaboration, partnership, engagement, involvement, and presence of Native people [at Hueco Tanks], that has a negative impact and effect on that place and on us every single day. Every day that there isn't a collaborative agreement and partnership, and engagement between us and that entity, Texas Parks and Wildlife, that's a negative thing. That's a bad thing. That place, it puts that place at risk. Every day that someone leaves there that has institutional memory and we get somebody fresh out of wherever that doesn't know, that doesn't have a relationship with Hueco—I'm not saying that that's inherently in and of itself a bad thing, but it can be a dangerous thing because they may not have that understanding, basic understanding of no, no, no, no, no, no. You don't go in there and try to manipulate the plants and the animals and manage yourselves. Manage yourselves. Be responsible for yourself and what you do. So I think if we did, if we implemented these, if we learned from Chaco Canyon, Canyon de Chelly, Uluru, we would create an even more powerful template and example.

[***RG***] *Should Hueco Tanks pursue recognition by UNESCO for Hueco Tanks as a World Heritage Site?*[19] *In your view,* *what advantages or disadvantages do such classifications present for the larger goals of preservation?*

[*Alex Mares*] Well, I think that the jury's still out as far as whether it would be a good thing or a bad thing for Hueco Tanks to be designated a UNESCO World Heritage Site. Is it worthy of it? This place is worthy as any place on the planet. Another important criteria question from UNESCO is, is there an outstanding universal value, for the place or the thing?. . . But one that's undeniable, and a lot of times it's not even real obvious, are the pictographs. [Most anthropologists] that have written about them have basically said, this may be the place of origin of the entire Katsina belief system and the Katsina imagery. That's a mouthful. That's like really heavy for us as Native people and in the Southwest. But then you expand it even further out. And I can't think of a single group of or community of human beings from the polar regions to the jungles, Africa, and the Amazon that cannot relate to faces or painted faces or masks or the use of masks. Every group understands those concepts or has a history of use or importance of those things. So to have a place like Hueco Tanks that has the largest collection of painted master painted faces in North America, possibly the world. That's powerful.

I want a good precedent. I want a precedent, a precedent that shows wide scale, broad community support, diversity, people of many different colors and many different backgrounds coming together to support this and do this. A diversity of tribes and Indigenous groups more than [the five currently recognized as having a special relationship to Hueco Tanks]. Come on. That would get behind it and support and be represented and would play a role. It would finally, really, truly improve and create a collaborative effort between TPWD and Native people and this community.

The risk, though, is if TPWD is not willing to finally address the gorilla sitting next to the elephant in the room, then it could . . . legitimize very damaging, incompatible, disrespectful recreational activities at a sacred place, at a sacred site. That's a dangerous precedent for me. So those are risks. I think of when I think of UNESCO, I think of the things and they come from the old Jewish faith: If not now, when? And if not me, who? But I think of it in the context, in its totality, if we don't finally come to terms and make a decision and stick by it and sever the ties with

> the sport, the sport of technical climbing and bouldering at sacred sites, if we don't do it now, when are we going to do it? We're never going to do it. And if we, us, all of us right now, living and breathing from the director of TPWD to the director of Texas Circle Mission to the governor, to everybody all the way down the line to UNESCO officials in Europe, if we don't do this, then who is? Who is going to do it?

Notes

1 This so-called "googly-eyed" Tlaloc figure (1 of over 20 such figures at Hueco Tanks) is typically assigned to the Jornada Mogollon period. See, in particular, Schaafsma, *Indian Rock Art*; Howard, et al., *10,000 Years at Hueco Tanks State Park and Historic Site.*
2 On the relation between pictographs and sovereignty, see esp. Garcia, *Signs of the Americas*, esp. 95–120.
3 Drawing on Walter Benjamin, Edgar Garcia speaks to these experiential dynamics through what he calls the "threshold magic" of pictographs, which he describes as being, "energetic, responsive, and indeed unnatural in its ability to continuously redefine the nature of the world," *Signs of the Americas*, xv.
4 Sherman, *Hueco Tanks*, 247.
5 Sherman, *Hueco Tanks*, 246–248.
6 As a gesture of respect, I choose not to reproduce examples of those names here. Here are the receipts: Sherman, *Hueco Tanks*, 390–407.
7 Sherman is himself one of the most legendary and revered figures in rock climbing and is the originator of the famous "V-rating," the globally recognized rating system for grading the difficulty of different bouldering problems (the "V" stems from his nickname, "Vermin"); that system was developed at Hueco Tanks. Author of multiple books and articles on climbing, Sherman is now the subject of a hagiographic documentary film. www.oldmanlightning.com.
8 Qtd. In Jackson, "Hueco Then and Now," 86.
9 Jackson, "Hueco Then and Now," 89.
10 Qtd. in Suchma, "Improvements at Hueco Tanks," 31.
11 See especially Glen Sean Coulthard (Dene), who has called for the rejection of the politics of recognition in favor of Indigenous models of sovereignty. *Red Skin, White Masks*, esp. chapter 1.
12 Qtd in Earle and Carmichael, "Rights of Passage," 5.
13 McCarthy, *Blood Meridian*, 173, 173.
14 McCarthy, *Blood Meridian*, 198.
15 McCarthy, *Blood Meridian*, 174.
16 Comar, "The Tigua Indians of Ysleta Del Sur."
17 Diamond, *Moon Spell.*
18 Garza, "Outdoor Recreation or Colonization?"
19 See Lynch, "From Handback to Landback."

Bibliography

Comar, Scott. "The Tigua Indians of Ysleta del Sur: A Borderlands Community." Ph.D. Diss. University of Texas at El Paso, 2015.

Coulthard, Glen Sean. *Red Skin, White Masks: Rejecting the Colonial Politics of Recognition.* Minneapolis: University of Minnesota Press, 2014.

Diamond, Tom. *Moon Spell.* Winston, NM: Beaver Lodge Press, 2015.
Earle, Duncan, and David Carmichael. "Rights of Passage: Cultural Claims to Hueco Tanks State Historical Park." *Practicing Anthropology* 27, no. 4 (Fall 2005): 5–9. https://www.jstor.org/stable/24781677.
Garcia, Edgar. *Signs of the Americas: A Poetics of Pictography, Hieroglyphs, and Khipu.* Chicago: University of Chicago Press, 2019.
Garza, Lilli Y. "Outdoor Recreation or Colonization?" *Drawstring Magazine*, June 26, 2021. https://www.drawstringmag.com/issue13/recreation-or-colonization.
Howard, Margaret, Logan McNatt, Terri Myers, Tim Roberts, and Amy Ringstaff, et al. *10,000 Years at Hueco Tanks State Park and Historic Site, El Paso County, TX.* Austin: Texas Parks and Wildlife Department, 2010.
Jackson, Jeff. "Hueco Then and Now: Six Players Speak Out." *Climbing Magazine*, June 2000. https://www.proquest.com/magazines/heuco-then-now-six-players-speak-out/docview/236411003/se-2.
Kish, Dawn, Dir. *Old Man Lightning.* http://www.oldmanlightning.com.
Lynch, Tom. "From Handback to Landback: Lessons from Uluru." In *Storied Deserts: Reimagining Global Arid Lands*, edited by Celina Osuna and Aidan Tynan. New York: Routledge, 2024.
McCarthy, Cormac. *Blood Meridian; or, the Evening Redness in the West.* New York: Vintage, 1985.
Schaafsma, Polly. *Indian Rock Art of the Southwest.* Santa Fe: School of American Research; Albuquerque: University of New Mexico Press, 1980.
Sherman, John. *Hueco Tanks: Climbing and Bouldering Guide.* 2nd ed. Evergreen, CO: Falcon Guides, 1995.
Suchma, Martin. "Off the Wall Access: Improvements at Hueco Tanks." *Climbing Magazine*, April 2004. https://www.proquest.com/magazines/improvements-at-hueco-tanks/docview/236453036/se-2?accountid=7121.

Index

Note: Page numbers in *italics* indicate figures; page numbers in **bold** indicate tables.

For Product Safety Concerns and Information please contact our EU
representative GPSR@taylorandfrancis.com
Taylor & Francis Verlag GmbH, Kaufingerstraße 24, 80331 München, Germany

www.ingramcontent.com/pod-product-compliance
Ingram Content Group UK Ltd.
Pitfield, Milton Keynes, MK11 3LW, UK
UKHW022316250926
475702UK00046B/892
9781032946757